Hydraulic Structures - Theory and Applications

Edited by Amimul Ahsan

Published in London, United Kingdom

IntechOpen

Supporting open minds since 2005

Hydraulic Structures - Theory and Applications
http://dx.doi.org/10.5772/intechopen.77330
Edited by Amimul Ahsan

Contributors
Costel Boariu, Adeniyi Ganiyu Adeogun, Abdul Rasak Apalando Mohammed, Raul Flores-Berrones, Amimul Ahsan, Monzur Imteaz, Musa Abubakar Tadda, Abubakar Shitu, Umar Abdulbaki Danhassan, Aliyu Idris Muhammad

First published in London, United Kingdom, 2020 by IntechOpen
IntechOpen is the global imprint of INTECHOPEN LIMITED, registered in England and Wales, registration number: 11086078, 7th floor, 10 Lower Thames Street, London, EC3R 6AF, United Kingdom
Printed in Croatia

British Library Cataloguing-in-Publication Data
A catalogue record for this book is available from the British Library

Additional hard and PDF copies can be obtained from orders@intechopen.com

Hydraulic Structures - Theory and Applications
Edited by Amimul Ahsan
p. cm.
Print ISBN 978-1-83881-995-8
Online ISBN 978-1-83881-996-5
eBook (PDF) ISBN 978-1-83881-997-2

Meet the editor

Associate Professor Amimul Ahsan was born in Netrokona, Bangladesh. He received a PhD in Civil Engineering from the University of Fukui, Japan. He has nearly 15 years' research, teaching, and industry experience. He has published extensively on water and environmental engineering, including nine books, 16 book chapters, and over 131 journal articles. He has received 14 international awards, including "Who's Who in the World 2015," "Leading Engineers of the World 2013," and the "Vice Chancellor Fellowship Award (Science and Technology)" from Sultan Selangor (Chancellor, UPM), Malaysia, in 2015. He is editor-in-chief of five journals in the United States, the United Kingdom, and Malaysia, and founder of the *Journal of Desalination and Water Purification* and the *Journal of Advanced Civil Engineering Practice and Research*. He is involved with several collaborative research projects globally and has a Scopus *h*-index of 22. He has been involved in organizing more than 40 international conferences as a session chair and co-chair, keynote speaker, invited/session speaker, and organizing/technical program committee (TPC) member in different countries for some time (e.g. 4th GCSTMR World Congress 2019, Dhaka; Global Civil Engineering Conf. (GCEC) 2017 at Malaysia; and 8th Innovation Arabia Conf. in February 16–18, 2015 at HBMSU, Dubai, UAE). He was a former faculty member of the Department of Civil Engineering and key researcher at the Institute of Advanced Technology (ITMA), University Putra Malaysia (UPM), Malaysia (2010–2017). Currently, he is an associate professor in the Department of Civil Engineering, Uttara University, Dhaka, Bangladesh, and adjunct associate professor in the Department of Civil and Construction Engineering, Faculty of Science, Engineering and Technology, Swinburne University of Technology, Melbourne, Australia.

Contents

Preface

Hydraulic structures such as dams, reservoirs, culverts, weirs, and spillways are engineering constructions designed and maintained for managing and utilizing water resources efficiently for the betterment of human beings and to save our environment. A dam is a very common structure, which is constructed across flowing water. It obstructs or directs or retards the water flow, commonly forming a reservoir. *Hydraulic Structures: Theory and Applications* conveys a broad understanding of the fundamental mechanisms of various hydraulic structures. Emphasis is given to the analysis and design of different types of hydraulic structures. Various applications of hydraulic structure analysis are also incorporated.

This book introduces advanced ideas on hydraulic structures: theory and applications to the international community. It includes five advanced and revised contributions, and covers (1) an introductory chapter on hydraulic structures, (2) the operation and maintenance of hydraulic structures, (3) a bottom discharge conduit for dams, (4) a review of methods of measuring streamflow using hydraulic structures, and (5) geotechnical engineering applied to earth-rock dams. The aim of the book is to provide a text for undergraduate and postgraduate students. Researchers, designers, and operators of hydraulic structures will find the text of interest and a stimulating up-to-date reference source.

Readers of this book will appreciate the current issues on hydraulic structures in different aspects. The approaches are applicable to various industrial purposes as well. The advanced ideas and information described here will be fruitful for readers to find a sustainable solution to problems encountered by an industrialized society.

The editor of this book would like to express sincere thanks to all authors for their high-quality contributions and in particular to the reviewers and assistants for reviewing/revising the chapters.

The editor would also like to express appreciation to all who have helped to prepare this book, especially Author Service Manager Ms. Dolores Kuzelj, Commissioning Editor Ms. Anja Filipovic and IntechOpen publisher.

Gratitude is expressed to Mrs. Ahsan, Ibrahim Bin Ahsan, Marium Binti Ahsan, Yahya Bin Ahsan, Yusuf Bin Ahsan, and my mother, father, mother-in-law, father-in-law, brothers, and sisters for their endless inspiration, mental support, and also necessary help whenever any difficulty arose.

Amimul Ahsan
Department of Civil Engineering,
Uttara University,
Dhaka, Bangladesh

Department of Civil and Construction Engineering,
Swinburne University of Technology,
Melbourne, Australia

Chapter 1

Introductory Chapter: Hydraulic Structures for Managing Water Resources Efficiently

Amimul Ahsan and Monzur Imteaz

1. General

This book (*Hydraulic Structures: Theory and Applications*) conveys a broad understanding of the fundamental mechanisms of various hydraulic structures. The emphasis is given on analysis and design of different types of hydraulic structures. Various applications of the hydraulic structures analysis are also incorporated in this book. The aim of the book is to provide a text for undergraduate and postgraduate students. Researchers, designers, and operators of hydraulic structures can find the text of interest and a stimulating up-to-date reference source.

2. Hydraulic structures

Hydraulic structures such as dam, reservoir, culvert, weir, and spillways are engineering constructions designed and maintained for managing and utilizing water resources efficiently for the betterment of human being and to save our environment. A dam is a barricade across flowing water that obstructs or directs or retards the water flow, commonly forming a reservoir [1–3]. **Figure 1** represents the typical forces acting on gravity dam. **Figure 2** shows the plan and longitudinal section of a rectangular culvert and a trapezoidal weir. **Figure 3** shows the hydraulic scheme and cross-section of a trapezoidal weir.

In general, a lake is an area that is filled with water and enclosed by soil, and it may have inlet(s) and outlet(s). An artificial lake can be a reservoir that is created

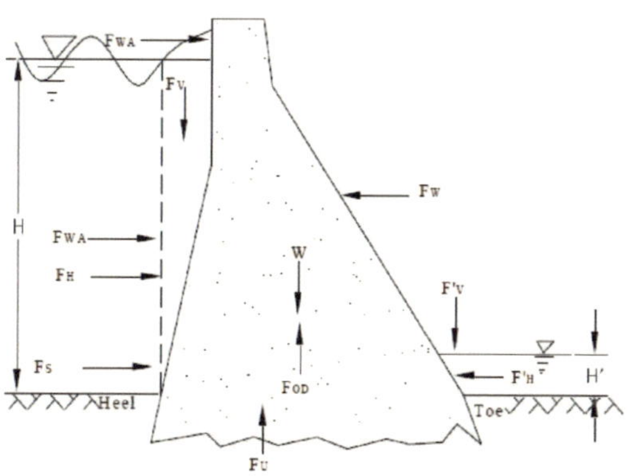

Where:
H = Head water depth
H' = Tail Water depth
F_{WA} = Wave pressure force
F_H = Horizontal hydrostatic force
F_S = Silt/sediment pressure force
F_{EQ} = Earthquake/Seismic force
F_W = Wind pressure force
F_H' = Tail water hydrostatic force
W = Weight of dam
F_{OD} = Internal pore water pressure
F_U = Uplift pressure force [base of dam]
F_V = Weight of water above dam [u/s]
F_V' = Weight of water above dam [d/s]

Figure 1.
Typical forces acting on gravity dam [1].

IntechOpen

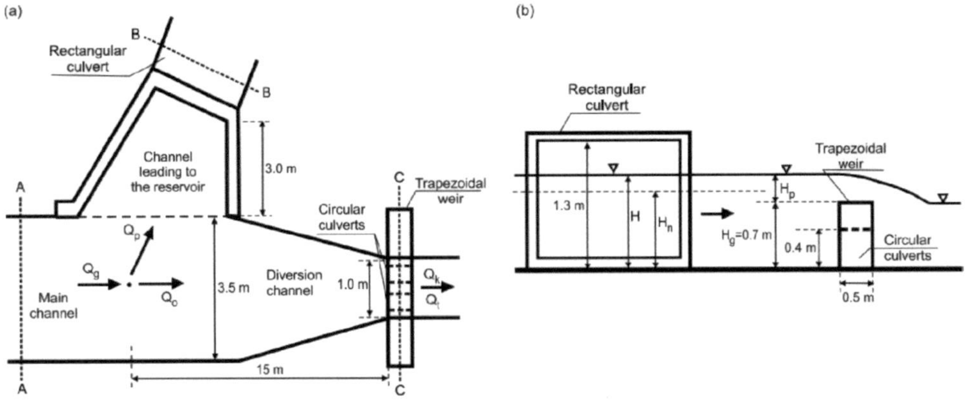

Figure 2.
Plan and longitudinal section of a rectangular culvert and a trapezoidal weir [4]. (a) Plan of culvert and weir, and (b) Longitudinal section of culvert and weir.

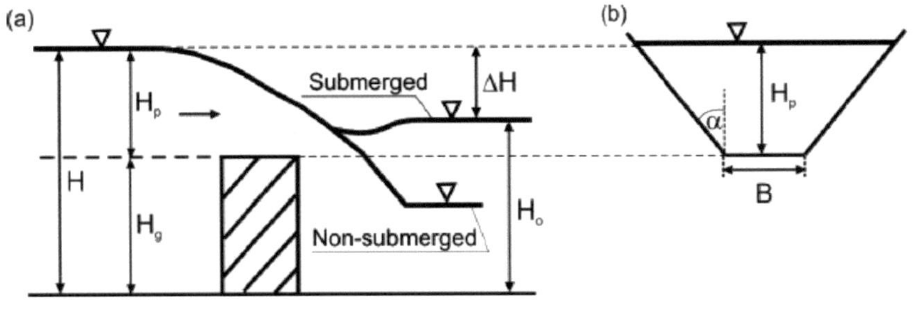

Figure 3.
Hydraulic scheme and cross-section of a trapezoidal weir [4]. (a) Hydraulic scheme of weir, and (b) Cross-section of weir.

behind an embankment or a dam by flooding soil. A few of the biggest lakes in the world are reservoirs. A spillway is a segment of an embankment or a dam intended to carry water downstream from the upstream side of an embankment or a dam. It may have doors that are intended to regulate the flow of water, i.e., flood. Flood can be defined as a water overflow on the soil, which is an accumulation of water over land that is not normally submerged [1].

The differences between dam and embankment and causeway are clarified. A dam is a wall constructed across a river to create a reservoir upstream side, an embankment is a built-up river or seawall (at the shore or bank), whereas a causeway is a high road or path across wet or tidal ground. A few railway embankments are also constructed using the same idea [3].

Ghomri et al. [5] studied the hydraulic jump of a hydraulic structure controlled by threshold, moving in a channel profile in a lab scale for a single roughness. The hydraulic jump was developed at the sharp transition from a supercritical flow.

3. Plant basket hydraulic structures (PBHS)

River restoration is attractive as it offers considerable benefits to the environment and economy. A new plant basket hydraulic structure (PBHS) as a new river restoration measure is applied in the Flinta River, central Poland. It focuses on changes of hydromorphological conditions in a small lowland river. This is a pilot project of the construction of vegetative sediment traps (plant basket hydraulic structure) [6].

4. Conclusions

Various types of hydraulic structures are used in the world. Each of them has pros and cons. The readers of this book (*Hydraulic Structures: Theory and Applications*) will appreciate the current issues on analysis of hydraulic structures in different aspects. The approaches would be applicable in various industrial purposes as well. The advanced idea and information described here on hydraulic structures will be fruitful for the readers to find a sustainable solution in an industrialized society.

Author details

Amimul Ahsan[1,2]* and Monzur Imteaz[2]

1 Department of Civil Engineering, Uttara University, Dhaka, Bangladesh

2 Department of Civil and Construction Engineering, Swinburne University of Technology, Melbourne, Australia

*Address all correspondence to: ashikcivil@yahoo.com

References

[1] Chane B, Behailu S. Hydraulic Structures I. Lecture Note. Addis Ababa, Ethiopia: Department of Civil Engineering, Faculty of Technology, Addis Ababa University; 2006

[2] Novak P, Moffat AIB, Nalluri C, Narayanan R. Hydraulic Structures. 4th ed. Oxon, UK: Taylor & Francis; 2007

[3] HiNative. What is the Difference Between Dam and Embankment and Causeway? Available from: https://hinative.com/en-US/questions/237962 [Accessed: 30 July 2019]

[4] Gąsiorowski D, Artichowicz W. Distribution of flows in a channel network under steady flow conditions. Acta Scientiarum Polonorum-Formatio Circumiectus. 2019;**18**(1):27-37. DOI: 10.15576/ASP.FC/2019.18.1.27

[5] Ghomri A, Riguet F, Debabeche M. Effect for a single roughness e=5,63mm of experimental to study hydraulic jump profile in a channel in u a rough bottom. Journal of Fundamental and Applied Sciences. 2015;**5**(1):25-39. DOI: 10.4314/jfas.v5i1.3. Available from: https://www.ajol.info/index.php/jfas/article/view/120940/110380

[6] Kałuża T, Radecki-Pawlik A, Szoszkiewicz K, Plesiński K, Radecki-Pawlik B, Laks I. Plant basket hydraulic structures (PBHS) as a new river restoration measure. Science of the Total Environment. 2018;**627**:245-255

Chapter 2

Operation and Maintenance of Hydraulic Structures

Musa Abubakar Tadda, Amimul Ahsan, Monzur Imteaz, Abubakar Shitu, Umar Abdulbaki Danhassan and Aliyu Idris Muhammad

Abstract

Water is among the most valuable resources that nature has endowed to human beings. Water has cut across all spans of life from cradle to grave. Since time immemorial, man continuously developed methods and techniques to harness the benefits of water and as well to protect himself from the destruction that may be caused by the same water. Therefore, for a hydraulic structure to answer its name, it must be capable of being used smoothly for the purposes it was designed for and also be able to be controlled effectively without the risk of causing any havoc to the environment. Using water, especially for agricultural purposes, cannot be overemphasized. Hence, this chapter discusses the hydraulic structures based on the work they performed, challenges facing hydraulic structures, and management procedures of the hydraulic structures in order to adequately and efficiently serve their purpose.

Keywords: hydraulic structures, operation, maintenance, water, design

1. Introduction

Hydraulic structures play an important role in drainage, irrigation, and hydraulic projects. If hydraulic structures fail, it may cause serious damages of wealth, properties, and environment as well as losses of life and injury to economy. The water related infrastructures are constructed at the aims to facilitate human needs/desires and enhance the quality of life such as drainage channel, river/channel, irrigation canal, bank/foot protection work, embankment, dam, spur dike/groyne, bridge/culvert, regulator, barrage/large regulator, aqueduct, pump station, siphon, and sluice. The details of some of the hydraulic structures are presented below.

1.1 Types of hydraulic structures

Hydraulic structures are structures that are fully or partially submerged in water. The essence of building hydraulic structures is to either divert, disrupt, store, or completely stop the natural flow of water bodies. Based on the work they are designed to perform on streamflow, hydraulic structures are categorized as water-retaining structures (dams and barrages), water-conveying structures (artificial channels), and special-purpose structures (structures for hydropower generation or inland waterways) [1].

1.1.1 Water-retaining structures

The dam is an essential hydraulic structure that all other structures directly or indirectly relied upon. Dams and barrages are typical *water-retaining structures* that are built purposely to impound water. The retained water behind dams and barrages could be used for other purposes such as irrigation, recreational activities, navigation, and a lot more. As of September 2019, there are 57,985 registered dams in the world [2]. Regardless of their size and type, dams demonstrate high complexity in their load response and interactive relationship with site hydrology and geology. Dams are of different sizes and shapes and made of various materials such as soil or rockfill embankment, mass concrete, reinforced concrete, masonry, and wood. However, based on the construction materials used, dams are broadly classified into concrete dams and embankment dams.

- Concrete dams comprised of gravity (PG), arch (VA), buttress (CB), barrage (BM), and multiple-arch dams (MV) as shown in **Figure 1a–e**. All these dams are constructed of mass concrete and sometimes of masonry with appropriate structural quality [1, 2]. Recent statistics show that concrete dams occupied only 20–22%, while embankment dams accounted for 78–80%.

- Embankment dams are of two types, earthfill (TE) and rockfill (ER), both of which are constructed by mass filling of naturally existing ground materials (soil and rocks). The construction materials are graded and well compacted to resist seepage and sliding. Embankment dams are characterized by having similar moderate face slopes at both upstream and downstream. This feature gives rise to a broad trapezoidal cross section and a high construction volume, which is relative to the dams' height that can cover >300 m [2].

1.1.2 Water-conveying structures

Any artificial facility cut in the ground with the sole purpose of transporting water diverted from main sources (river and dams) is termed as the *water-conveying structure*. These types of structures are comprised of canals (**Figure 2a**) and tunnels (**Figure 2b**) (usually made from soil and rocks) or siphons, aqueducts (**Figure 2c**), flumes (**Figure 2d**), and pipelines (usually made from concrete and metals) [1]. Before the construction of any water-conveying structure, a detailed geotechnical soil test and analysis is recommended to avail the surface and subsurface properties of the soil on which the structure is upon rest. The same soil test and analysis also applies to other types and classes of hydraulic structures to ensure safety and to save resources.

1.1.3 Special-purpose hydraulic structures

As the name implies, *special-purpose hydraulic structures* are built as an integral part of hydraulic project to meet a special purpose such as hydropower generation (e.g., surge towers and shafts, forebays, and head ponds), navigation (e.g., landings, berths, substations for ship repair, etc.), fishing (e.g., fish nursery ponds, fish lifts and locks, fishways, etc.), water supply for domestic and industrial uses (e.g., water intakes to treatment plant, pumping stations, etc.), waste disposal/sewerage (e.g., sewage headers, pumping stations, channels after treatment plant to water bodies, etc.), and land reclamation (e.g., irrigation canals, drainage systems, silt tanks, etc.) [1, 7].

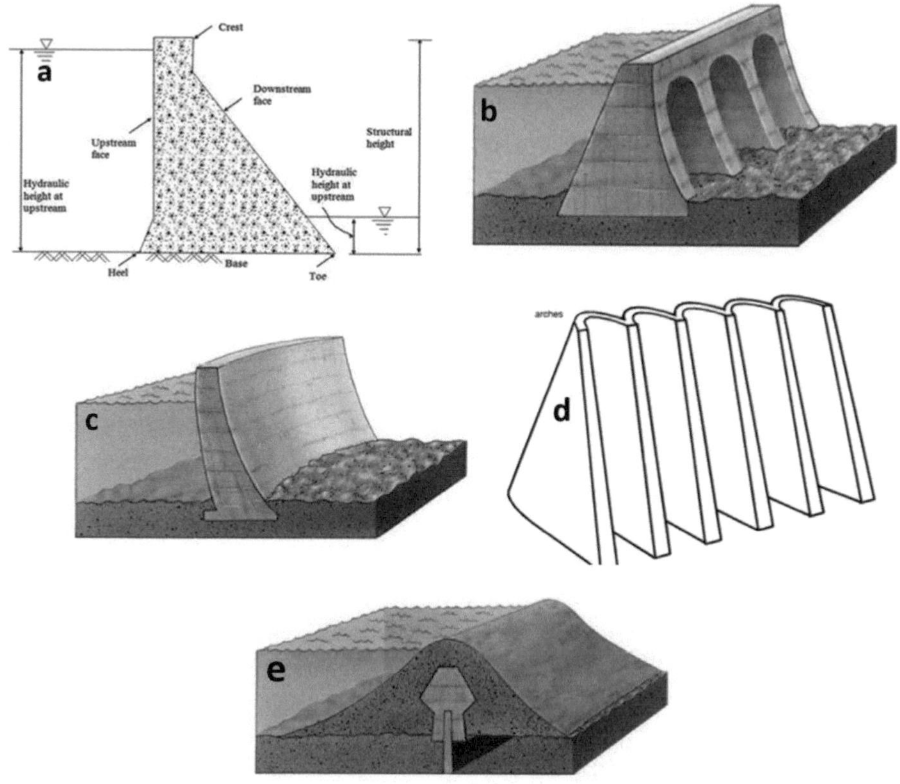

Figure 1.
(a) Gravity dam, (b) arch dam, (c) buttress dam, (d) multiple-arch dam, (e) earthfill and rockfill dam. Source: [3].

Figure 2.
(a) Canals, (b) tunnels, (c) aqueducts, and (d) flumes. Source: [4–6].

2. Purposes of hydraulic structures

Hydraulic structures are purposely for managing and controlling the flow of water in natural and built environment systems. Moreover, the primary purposes may include the following flood control, water conveyance, irrigation, navigation, power generation, domestic and industrial purposes, environment protection, and recreation, among others.

2.1 Flood control

Flooding is a geophysical hazard that nonuniformly dispersed in both space and time. Over a decade, several watershed areas are frequently suffering from flood disaster, which causes massive destruction and loss of lives, farmlands, crops, access roads, and houses [8]. The effective way of flood control and reducing its negative impacts is by the construction of dams, water conveyance structures, culverts, canals, and reservoirs [9]. Many control structures are not solely constructed mainly for dealing with flood control only. However, sometimes, hydraulic structures are purposely built for flood control only. In the designing and building of flood control structures, some vital point of views must be taken into consideration in such that the cost of construction of such a project structure should be of benefit, concerning the damage reduction and the public interest when comparing to similar benefits to be derived by the alternative means. Also, the flood control structures should be reliable and effective as predicted. Even in some instance, the methods of controlling floods should rather be automatic, not manual.

2.2 Power generation

Hydropower generation is the production of electrical energy from running water through turbines without reducing its quantity. The flexibility; long-lasting, storing capability; less environmental pollution; and the cost-effectiveness of hydropower plants make it attract more investment as a renewable energy source and role as a way of drought mitigation [10]. It has been demonstrated that hydropower generated about 16.4% of the global total electricity supply equivalent to the installed capacity of about 1064 GW [9]. The hydropower system is the leading global source of an estimated 71% of total renewable energy. Furthermore, hydropower plant reservoirs can also be used as a tool in minimizing the adverse impacts of climate change and in achieving sustainable development goals [11].

2.3 Navigation

Inland water transportation plays a significant role in the national and global markets. Building dams and draining of river streams will considerably raise the capacity of inland water transportation, thereby allowing the smooth movement of a shipping vessel. An important point to note is that a chain of storage reservoirs would advance navigation depth, straightening out navigation channels, and support the passage of both small, medium, and even large ships. However, it is recommended to provide pathways or locks for vessels when dam structures are built on a large river stream for easy navigation from upstream to the downstream. Also, the topography of the surrounding environment should be taken into consideration. Hence, the pathways might be an integral part of the dam or a completely different structure.

2.4 Irrigation schemes

Recently, it was reported that about 20% of the global total arable land is under different forms of irrigation schemes. More than 70% of freshwater withdrawn from rivers is utilized for irrigating crops, and 75% of the total water hardly returns to the rivers [1]. In many regions of the world, with water scarcity, farming without irrigation would not be possible. The quantity of water kept in the storage reservoirs and the power required for water pumping are provided by hydropower plants, which are integral parts of the multipurpose hydraulic structure. In the present world, irrigation projects depend on the supply from multipurpose hydraulic dams, reservoirs, and rivers. For irrigation schemes to be successful, the water supply from sources must be adequately available whenever needed and at a reasonable cost of investment. Also, the operation and maintenance of such a structure should be smooth and cost-effective.

2.5 Municipal and industrial water supply

A large quantity of freshwater is being consumed daily by food processing; mineral mining and processing; textile, paper, and pulps; nuclear and thermal power plants; and drugs and pharmaceutical, petrochemical, and metallurgical industries, among others. However, some of the major industries that use a large volume of water are nuclear and thermal power plants. To meet both domestic and industrial needs, due to the higher demand for water by many industries, especially in industrially developed nations, large capacity storage structures are always built to store local rainfall runoff and water diverted from other river basins. Multipurpose hydraulic structures are the primary storage and sources of most water supply for domestic and industrial purposes. Although public water consumption constitutes nearly only 10% of the water consumed by the industries, still the immediate needs of public water supply must be taken seriously [12]. The water supply from hydraulic projects should always meet the standards of quality required for domestic and industrial uses in terms of its color, test, hardness, odor, and bacterial purity. Also, the treatment methods for the water should be cost-effective and daily available all year round. Necessary control and protection measures should be provided in the river basin areas where the hydraulic project is sited which are mainly for the municipal water supply. The need for hydraulic projects is also in a region with the seasonal variation of rainfall distribution of the year.

2.6 Environment protection

Another vital reason for hydraulic projects is for environmental protection and water management, which may include farmland improvement by controlling soil erosion; environmentally friendly hydropower supply; improved quality water supply for human, animal, and industrial consumption; aquatic food supply; and recreational development [8]. Nevertheless, the negative impacts posed by the massive hydraulic structures on the environment and public safety should always be considered in the course of design and construction processes [1]. The essential environmental issues are for the well-being of people living around the hydraulic projects and to the other plants and animals for the social needs of humankind.

2.7 Recreation and other purposes

Many hydraulic projects also serve as a place for tourism, recreational, and sports activities. In fact, in some countries, sometimes hydraulic projects are specially constructed for recreation purposes. Some recreational activities carried out at the hydraulic project sites might include swimming, fishing, boating, canoeing, scuba diving, and lakeside walking. Recreational activities provide job opportunities to the teeming population and generate incomes to the government and, at the same time, conserve the natural environment.

3. Operation and maintenance of hydraulic structures

Strategies for sustainable operation and maintenance of hydraulic structures are initiated before design and are optimized during its service life for the safety of lives and properties, which stabilizes the environment and the national economy. Consequently, improper hydraulic structures' operation and maintenance may lead to loss of life, properties, economy, and the environment. The responsibilities for the operation and maintenance of hydraulic systems vary in different countries, depending on the ownership and purposes. In Nigeria, the responsibilities rest on the central government, coordinated by the department of water resources. This section has highlighted the necessary strategies for safe operation, maintenance, and consequences due to failure. The strategies can be long term, seasonal, frequent, and daily. The primary tasks to exemplary operation and maintenance of hydraulic structures according to Chen [1] are as follows: hydrologic monitoring and forecasting, detection and mitigation of aging of structures, safety surveillance and instrumentations, and remedial actions.

3.1 Hydrologic monitoring and forecasting

Safe operation and management of hydraulic structure primarily depend on the efficiency of metrological stations to provide independent data of water regime and observation. The data obtained can be used during the analysis and prediction of future hydrologic events. Nowadays, automated facilities are installed at various locations in the catchment area to provide hydrologic data. After the analysis of the data, the forecasted value and period must be provided with some reliable accuracy. The short-term forecasting, developed on runoff and other fundamental theories, provides the basis of flood controls in the catchment. Mid- and long-term forecasting give essential information to the hydropower sector [1].

3.2 Safety surveillance and instrumentations

The continuous, systematic assessments of the physical condition of hydraulic structures without compromise are encouraged. The large capacity hydraulic structures constitute a more significant threat to downstream life and properties. Mostly, failure arises from extreme flood events and inter- or obvious structural distress, which necessitates safety surveillance and instrumentation programs to detect the possible symptom and specific problem at an early stage in hydraulic structures and create strategies for the solution to the possible abnormalities [1, 13]. The selection and installation of equipment or instrumentation at appropriate locations in the surveillance area, adequate interpretation of the surveillance data, and immediate actions are more important than the number of instruments installed.

3.2.1 Safety inspection

The safety inspection is a regular inspection of some deteriorations to determine the current state of hydraulic structures based on purposes related to the operation.

Safety inspections are categorized into routine, specialized, and periodic inspections. Specifically, the embankments of large capacity structures should be closely and routinely examined against any physical defect [13]. This inspection is categorized into routine, specialized, and periodic inspections [1], and thus, their cumulative records determine whether a defect is new, gradual, and/or rapidly changing in the structures [13]. The routine inspection aims to identify the physical deficiencies of the hydraulic structures during day-to-day operations. Periodic inspections are carried out by experienced technical crews at an interval of 2–3 years and are meant to detect physical defects on the structures by visual examination so that strategic remedial actions can be taken. Specialized inspections include earthquake and check-flood inspections. Earthquake and check-flood are identified as a potential threat to hydraulic structures. Their inspection is carried out by experienced and well-trained dam engineers. Thus, the documented reports for mitigations are then put into the remedial action plan.

3.2.2 Surveillance and instrumentation

Surveillance is the continuous monitoring of physical conditions through medium to large instruments. It is being done to check the deterioration concerning the actual performance of the hydraulic structure and its trends for compliance with the design expectations. In this operation, the collection, presentation, and evaluation of data from the equipment installed in the system are paramount. The equipment must cover critical components and should be installed at positions where normal behavior is anticipated. It is a good practice to draft an ideal instrumentational plan at an early stage to eliminate the less essential provisions until an adequate, balanced, and affordable plan is determined. In large-scale structures such as a dam, surveillance through high-level technology should be enhanced. Monitoring of change in temperature and cracks occurring in the embankments are used to reveal seepage and sediments during operations.

3.3 Remedial actions

Remedial actions are meant to prevent failures of hydraulic structures, especially the large capacity structures that pose a significant threat to lives and properties. The deficiencies are classified as minor, moderate, and major accidents [1]. Their remedial actions are necessary before the failure of the entire structure. The defects may earlier be detected through surveillance, and the defects may probably be design-related, such as improper design capacity, or construction-related such as inappropriate choice of materials. The common and challenging operation- and maintenance-related incidents are the rapid rises in seepage, overtopping of earth embankment, excessive beaching, erosion of spillway and embankments, cracking in the concrete dam and spillway, and fractured gates. The remedial actions to be considered depend on the condition of structures and hydrologic events. The remedial measures included:

1. Preventive control to reduce the condition from escalation

2. Short-term actions to modify the nearby catchment conditions, such as increasing surveillance, emergency evacuations, and lowering the overtopping

3. Long-term remedies in the structures, such as reinforcements, gates, dredging, and abdication

3.3.1 Emergency remedial actions

a. *Erosion control:* During floods, the use of polyethylene sheeting and sandbag controls the erosion of the slope embankment [1].

b. *Overtopping control:* Overtopping must be avoided, and the provision of temporary barrier above the predicted altitude is applied.

c. *Seepage control:* The seepage must not be allowed to saturate the downstream slope, and if saturated, the provision of permeable material to reduce pressure buildup on the embankment is needed.

3.4 Detection and mitigation of aging

3.4.1 Aging of hydraulic structures

Aging of a hydraulic structure is referring to the time-related deformations in the properties of the material and its foundation used during construction of the hydraulic structures, which developed within at least 5 years of working period. Also, it is the entire lifespan of hydraulic structure before abdication or decommissions. The deterioration of the structures may be due to the defects developed through unusual events such as an earthquake or a result of environmental factors during service life.

3.4.2 Detection of aging

Detection of aging should start during the operation and maintenance of hydraulic structures. Factors that influence the degradation of the structural properties of hydraulic systems should be identified and must immediately be managed. Alternatively, nondestructive examinations could be essential to detect the aging of hydraulic structures. The nondestructive examinations are the direct and indirect evaluation of information regarding the state of the hydraulic structure. This is to allow for immediate interventions in the situation and avoid severe consequences. Indirect assessment of aging should be accomplished by monitoring the effects and consequences of aging.

On the other hand, the direct assessment is performed by inspecting and testing the data of the structural properties of the hydraulic structures. The laboratory experiments and the in situ assessments, where the physical and mechanical properties of the sediment, including concrete, are extracted and analyzed, are examples of destructive examination. According to Chen [1], the destructive examination with in situ tests may or may not be destructive. The destructive examinations may include (i) hydraulic pumping tests for porosity and (ii) permeability and leak detection through a physical and chemical test of catchment and leakage, among others.

Similarly, a nondestructive examination is designed to ascertain the flows of materials while it protects the object's usability, successfully nondestructive tests, and requires an understanding of its limitations and data manipulation. Various methods, such as electromagnetic, resistivity, acoustic, induce polarization, and visual assessment, are employed.

3.4.3 Mitigations for aging

Adequate mitigations of aging of hydraulic structures start during the designs, effected during construction, which continues through monitoring and surveillance in operation and maintenance stages. The prior understanding of the factors that influence the degradation of the structural properties of the materials used in the constructions of the hydraulic structures must be scrutinized. Also, the provision of extra quality to meet the designed lifespan of the system must be put into consideration during the constructions. Alternatively, the following mitigations steps are commendable:

a. *Analysis:* The analysis of the aging process is carried out to ascertain its severity to the safety of life, properties, national economy, and environment.

b. *Prevention:* It is well known that all structural materials have a finite lifespan and can be affected by the environment. The prevention stage to mitigate aging of a hydraulic structure is proceeded by detailed analysis to know the structure's safety and its economic condition. If the effect is infinite, immediate remedial action such as an emergency action plan is necessary. However, if the effect is finite, and the structure has an economic lifespan, then, provision of concrete structures from uniquely selected materials is encouraged.

c. *Rehabilitations:* Many physical and chemical methods like geomembrane are employed to enhance waterproof. Additionally, the repair and replacement of corroded steels and the use of excellent impermeable materials are also administered for overlay operations.

4. Challenges facing hydraulic structures

The importance of hydraulic structures cannot be overemphasized, and therefore their maintenance and safe utilization are critical. The structures should neither leak nor erode; channels and structures should be clean and free from siltation with noncorrosive or rotten moving parts. The breakdown or failure of these hydraulic structures can lead to a disastrous situation within the surrounding areas. For instance, a catastrophic dam collapse could lead to flooding and erosion.

The challenges of maintaining hydraulic structures at the initial stage can be achieved by managing the characteristic of the flow to meet the desired goal of the project needs. According to Chen [1], this can be realized by considering the public safety and ecological, environmental, and the design objectives of each structure. Some of the challenges facing hydraulic structures and the way they can be addressed are further discussed in the subsequent section.

4.1 Erosion

Soil is a nonrenewable resource that supports human and animal life. Soil provides living beings with food, fiber, and protection from harsh environmental conditions such as high temperatures and heavy rainfall. Soil is lost due to erosion as a result of continuous cultivation of land, drastic reduction in vegetation, and collapsing of hydraulic structures such as dams. Erosion is the washing away of the topmost soil layer by the agents of erosion, including water, wind, and human activities [14]. Erosion by water is caused by overland flow and transport of sediments due to the interactive action of water flow and heavy rain droplets. In hydraulic structures, erosion can occur in canals, for example, in an unlined canal at downstream or lined canal section that receives water jet flow from a gate or pipe or water that spills over a weir. This type of erosion can be remediated by dissipating the energy of the incoming water through the construction of a stilling basin as part of the hydraulic structure immediately downstream of the pipe or weir [15]. Another critical point of canals that is prone to erosion is the intersection of a lined and unlined canal, that is, the transition point from a lined canal to the unlined canal, as shown in **Figure 3**. This type of problem is called undermining and, if not taken care of, can cause a collapse of the lining and destruction of the structure [16]. So, periodic maintenance should be observed to solve this problem. Undermining can be avoided or controlled by the provision of cutoff that will protect the foundation of the structure.

4.2 Leakage

Leakage in hydraulic structures refers to the ability of confined or upstream water bodies to exploit the least available exit, space, or crack underneath or along the structure to escape to the downstream or unconfined surrounding area. The moment the water found these small spaces, then there is a leakage problem, which is the beginning of erosion in the area. These small openings and cracks are widened with time and intensity of leakage. Thus, the soil is washed away as time goes on and the structure will collapse. At this point, preventing the collapse of such a structure will be very difficult. Take a dam, for instance, the water level is very high at the upstream. Water can flow along the dam embankment; if no measure is taken to save the structure, it can be undermined and collapse due to erosion [17].

It has been recommended by van den Bosch and Snellen [16] to observe and identify leakages at their initial stage and correct them. Leakages in the crack can be repaired by cleaning the wall or the floor where the crack is located. Then remove any sand, clay, plant growth, or debris. Open up the crack to become broader and more in-depth. Prepare cement-sand mortar to fill the hole and smoothen it with a trowel. Provide adequate curing to the repaired crack.

On the other hand, vertical cutoffs can be constructed on the structures to obstruct the flow of water underneath and along with the structure. An example of a cutoff wall in a dam is showcased in **Figure 4a**. Similarly, drop structures can also be equipped with cutoffs to block the water flow along and underneath the structure (**Figure 4b**). The cutoffs are part of the structure, driven into the embankments of a canal by digging deep into the banks of the canal and canal bed. During the installation, the earth around the canal banks and the cutoffs must be well compacted.

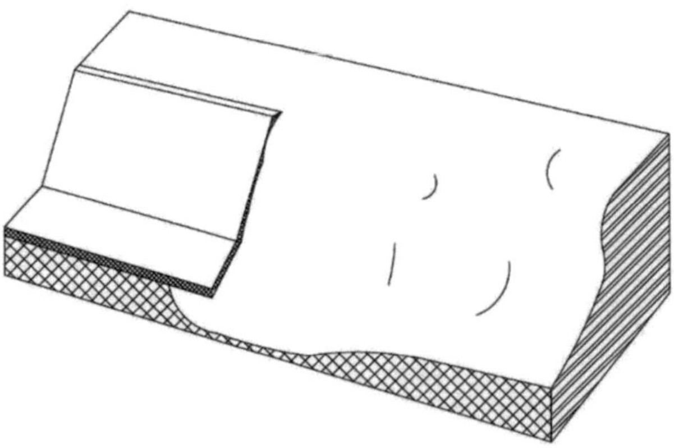

Figure 3.
Points of transition between a lined and unlined canal.

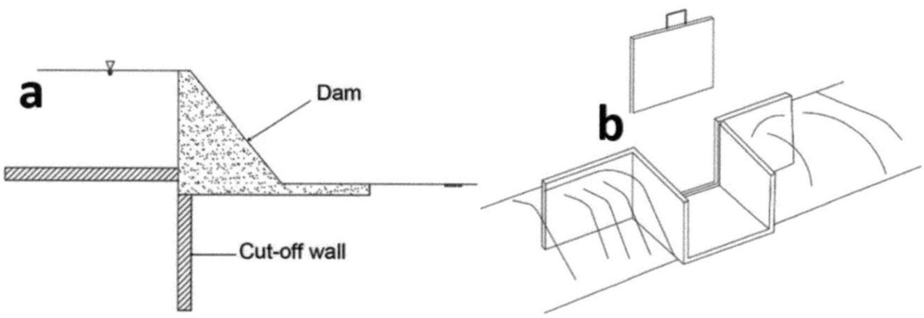

Figure 4.
(a) The function of the cutoff here was to prevent piping failure and reduce leakage or seepage. The cutoff was constructed parallel to the centerline of the dam (b) intake structure provided with a concrete cutoff wall.

4.3 Siltation

Siltation is the process of deposition of debris and sand particles and their buildup in hydraulic structures that obstruct the full functioning of the structures. The problems caused by siltation are usually the changes in water flow, changes in velocities and water levels, decreased energy dissipation, and so on. Examples of these problems include deposition of large volumes of sand in the intake chamber of pumps, which usually causes damage to the pumps and subsequent silting of the canals by sand particles. Another instance is siltation at the stilling basin. This type of sand deposits reduces the energy dissipation of the structure. Similarly, the changes in flow and velocities of water inflow division box are affected by sand particles deposited in the structure [16]. Because of these problems, large sand traps are usually constructed at the end of the upper main canal to collect the sand deposits and remove them by periodic cleaning.

4.4 Corrosion/rot

Hydraulic structures are made from different materials, including concrete, wood, or steel. These structures are liable to deterioration with time and with alternating wet and dry conditions subjected. The wooden parts in the structure, for instance, rot and decompose, whereas the steel parts corrode, as a rule, causing their expansion, and get jammed in the sliding slots. Such a condition affects the smooth operation of the structures. Routine maintenance is necessary to curtail the problems and reduce their effects. Painting of the affected parts can preserve them against corrosion. Lubrication of moving parts (steel) such as sluice gates and valves can prevent jamming.

Author details

Musa Abubakar Tadda[1,2], Amimul Ahsan[3,4*], Monzur Imteaz[4], Abubakar Shitu[1,2], Umar Abdulbaki Danhassan[1,5] and Aliyu Idris Muhammad[1,2]

1 College of Biosystems Engineering and Food Science, Zhejiang University, Hangzhou, China

2 Department of Agricultural and Environmental Engineering, Faculty of Engineering, Bayero University, Kano, Nigeria

3 Department of Civil Engineering, Uttara University, Dhaka, Bangladesh

4 Department of Civil and Construction Engineering, Faculty of Science, Engineering and Technology, Swinburne University of Technology, Melbourne, Australia

5 Department of Agricultural and Bio-environmental Engineering, SCA/DAC, Ahmadu Bello University, Zaria, Nigeria

*Address all correspondence to: ashikcivil@yahoo.com

References

[1] Chen S. Hydraulic Structures, First edit. London: Springer; 2015

[2] ICOLD. 2019 Data. International Commission on Large Dams [Online]. 2019. Available from: https://www.icold-cigb.org/GB/world_register/general_synthesis.asp

[3] How it Works: Dam Engineering [Online]. 2013. Available from: https://www.howitworksdaily.com/dam-engineering/ [Accessed: 07 October 2019]

[4] Panoramio [Online]. 2019. Available from: http://www.panoramio.com/photo/14830796 [Accessed: 07 October 2019]

[5] JoAnna W. Ancient Roman Aqueducts Could Spill Climate Secrets. Earth and Space Science [Online]. 2015. Available from: https://eos.org/articles/ancient-roman-aqueducts-could-spill-climate-secrets [Accessed: 07 October 2019]

[6] OCF. Fiberglass Parshall Flumes at Wastewater Plants. Open Channel Flow [Online]. 2019. Available from: https://www.openchannelflow.com/blog/fiberglass-parshall-flumes-at-wastewater-plants [Accessed: 07 October 2019]

[7] UDFCD. Urban Storm Drainage Criteria Manual: Hydraulic Structures. USA: Colorado; 2017

[8] Yazdi J, Salehi Neyshabouri SAA. Optimal design of flood-control multi-reservoir system on a watershed scale. Natural Hazards. 2012;**63**(2):629-646

[9] Nguyen-Tien V, Elliott RJR, Strobl EA. Hydropower generation, flood control and dam cascades: A national assessment for Vietnam. Journal of Hydrology. 2018;**560**:109-126

[10] Chilkoti V, Bolisetti T, Balachandar R. Climate change impact assessment on hydropower generation using multi-model climate ensemble. Renewable Energy. 2017;**109**:510-517

[11] Hasan MM, Wyseure G. Impact of climate change on hydropower generation in Rio Jubones Basin, Ecuador. Water Science and Engineering. 2018;**11**(2):157-166

[12] Kucukali S. Municipal water supply dams as a source of small hydropower in Turkey. Renewable Energy. 2010;**35**(9):2001-2007

[13] Zhu P, Leng YB, Zhou Y, Jiang GL. Safety inspection strategy for earth embankment dams using fully distributed sensing. Procedia Engineering. 2011;**8**:520-526

[14] van Lier HN, Pereira LS, Steiner FR. CIGR Handbook of Agricultural Engineering, First., Vol. I, No. 2. United States of America: American Society of Agricultural Engineers; 1999

[15] Hager WH, Boes RM. Hydraulic structures: A positive outlook into the future. Journal of Hydraulic Research. 2014;**52**(3):299-310

[16] van den Bosch BE, Snellen WB. Structures for Water Control and Distribution. Rome: Food & Agriculture Organisations - Technology & Engineering; 1993

[17] Paxson GS, McCann MWJ, Landis ME. A risk based framework for evaluating gated spillway operations. In: Hydraulic Structures and Water System Management. 6th IAHR International Symposium on Hydraulic Structures; Portland, OR; 27-30 June, 2016. pp. 630-640

Chapter 3

Bottom Discharge Conduit for Embankment Dams

Costel Boariu

Abstract

The structural calculation methods of the conduits that cross the embankment dams can be divided into two approaches. On the one hand, the estimation of the earth's characteristics is done by the multiparameter subgrade model, and on the other hand, the usual finite element software describes the parameters of the earth based on the modulus of elasticity. The conduits have a moment of inertia value for the cross section that is combined with the modulus of elasticity of the material (concrete, steel) resulting in a great rigidity to the assembly. This stiffness, quantified in the stiffness matrix, is much higher than the rigidity of the soil. This work goes in two directions: on the one hand, it argues that the complex methods of calculating the soil characteristics are not relevant for conduits that cross the embankment dams. Second in the longitudinal direction, the conduit has joints that diminish the rigidity of the assembly and whose effect cannot be included in the FEM calculation, as it is usually done in a plane strain model. A calculation method is proposed that contains an inertial moment adjustment that takes into account joints. Finally, a computational method is used in which FEM is used with the empirically estimated momentum variation.

Keywords: bottom discharge conduit, embankment dam, soil-structure interaction, finite element method, multiple-parameter subgrade model, matrix condensation

1. Introduction

Bottom discharge conduits are pipes that cross the body of the dam from the upstream to the downstream. Conduits can be made of reinforced concrete or metal. Less rigid materials can be used for small dams (PEHD, GRP-glass reinforced plastic). Passing a conduit through the body of an embankment dam or beneath its foundation requires a lot of caution and adequate construction measures [1].

The contact surface between the pipe and the embankment is a possible way for infiltration. These are prevented by special shoulders that increase the water infiltration path.

The modeling of the interaction between the structure and the earth filler must solve the following aspects [1]:

a. Calculating the pressures of the embankment on the conduit

b. Assessment of the soil (subgrade) reaction

IntechOpen

c. Calculating the displacements and deformations of the structure

d. Evaluating the state of stress of the cross section and longitudinal section

e. Cross-sectional and longitudinal cross-sectional composition (joints)

Due to the complexity of the interaction among the structure, the foundation, and the filling, the only current method able to accurately model this phenomenon is the finite element method (FEM).

For the first aspect, (a) the Marston theory of embankment pressures has typically been adopted for calculating loads on a conduit that is partially or fully projecting above the original ground surface [2–4]. Using the Marston theory, vertical load on the conduit is considered to be a combination of the weight of the fill directly above the conduit and the frictional forces, acting either upward or downward, from the adjacent fill.

For aspect (b) assessment of the ground reaction, there are calculation methods in which the behavior of the earth is modeled by springs with behavior not necessarily linearly elastic with or without interaction between them (between springs) [5]. These models introduce more parameters between the linear stiffness of the soil layers and the shear layers. Later [6, 7] the interaction between the structure and the terrain was described by the beam-column analogy as it incorporates the lowest level multiple-parameter subgrade model possible.

For aspects (c) and (d), guidance on the design and construction of conduits are provided in [2, 3, 8, 9]. The references contain accepted methods to design conduits. Reinforced concrete conduits are used for medium and large dams, and precast pipes are used for small dams, urban levees, and other levees where public safety is at risk or substantial property damage could occur [2]. Corrugated metal pipes are acceptable through agricultural levees where the conduit diameter is 900 mm (36 in.) and when levee embankments are no higher than 4 m (12 ft) above the conduit invert. Inlet structures, intake towers, gate wells, and outlet structures should be constructed of cast-in-place reinforced concrete. However, precast concrete or corrugated metal structures may be used in agricultural and rural levees.

Conduit composition in cross-sectional and longitudinal is detailed in [1, 10]. The conduits are made from 10 to 12-m-long sections in order to be able to adapt without cracking with the eventual differentiation of the foundation ground settlement. The outer shape of the cross sections should consider the interaction of the structure with the filler. Curved vault sections are the most recommended. Rectangular outer sections contour lead to stress concentrations and may lead to cracking of the sealing core along the structure.

Next we aim to find the importance of the assessment of the soil (subgrade) reaction. There are many methods of calculating the soil characteristics [5–8]. Which one is suitable for this type of structure? We propose to answer this question.

2. Interaction between soil and conduit

Most structural computing software currently used (e.g., SAP 2000 [11]) included only the spring stiffness connection for the ground displacements similar to the Winkler environment. Computational programs with geotechnical specialization allow the adoption of more complex models of behavior for the earth. However, besides the inability to capture the interaction between the structure and the ground, another deficiency results from the fact that the loads are obtained by a structural horse considering the fixed supports [5].

Everyone agrees that MEF is the best (exact) calculation method. How do we define the soil in the calculation model?

Most software models the ground through the Winkler springs. Is it accurate enough, or do we need more sophisticated methods of assessing the ground characteristics of the foundation? There are soil-modeling methods [5–7] that quantify the Winkler spring stiffness and a shear effect in the ground.

There are finite element programs (GeoStudio [12]) that model the behavior of the earth using for soil stiffness—Young's modulus (E). This modulus, which represents the stiffness of the soil, is dependent on the effective confining stress. In FEM (finite element method), finally, an equation is obtained that has unknown displacements and whose terms are stiffness matrices and loads.

Both the use of the multiparameter model subgrade and the use of the modulus of elasticity as a function of stress represent approximations of the effective situation.

In order to evaluate these calculation methods, the structure and earth influence on the rigidity matrix are calculated next. Considering the pipe geometry (large length element), it is possible to approximate the pipe with a beam.

2.1 Stiffness matrix calculation for continuous support

In the traditional method for simulation, the mathematical load-deformation response of a beam in uniaxial bending is a differential equation [6]. The basic form of the matrix formulation for beam flexure is (Eq. (1)):

$$[S]\{d\} = \{q\} \tag{1}$$

where $[S]$ is the stiffness matrix; $\{d\}$ is the displacement vector; $\{q\}$ is the load (force) vector.

The relevance of (Eq. (1)) is that all of the variations in beam behavior can be explained as variations solely in the formulation of the stiffness matrix, $[S]$.

In the Winkler model (**Figure 1**), the flexural behavior of this beam is given by Eq. (2):

$$EI\frac{d^4w(x)}{dx^4} + p(x) = q(x) \tag{2}$$

where subgrade reaction in one (x-axis) direction only is

$$p(x) = k_w w(x)$$

k_w is the Winkler coefficient of subgrade reaction; E is the modulus of elasticity (Young's modulus); I is the moment of inertia of beam section.

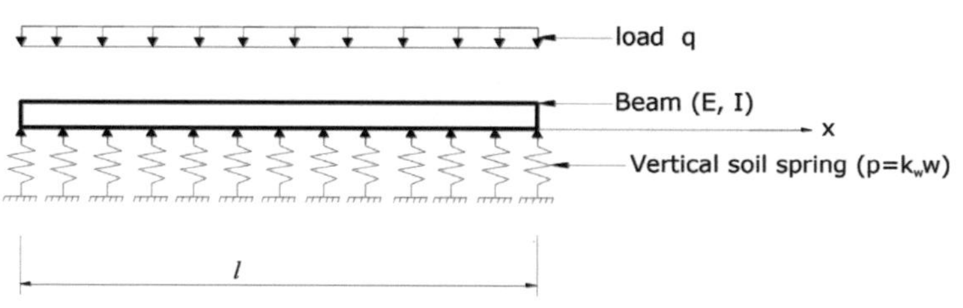

Figure 1.
The Winkler model.

Solving Eq. (2) by FEM is expressed by Eq. (3):

$$([S_e] + [S_w])\{d\} = \{q\} \tag{3}$$

in which the expressions of the elastic stiffness matrix of the beam $[S_e]$ and subgrade reaction matrix $[S_w]$ are [5].

$$[S_e] = \frac{EI}{l^3}\begin{bmatrix} 12 & 6l & -12 & 6l \\ 6l & 4l^2 & -6l & 2l^2 \\ -12 & -6l & 12 & -6l \\ 6l & 2l^2 & -6l & 4l^2 \end{bmatrix} \tag{4}$$

$$[S_w] = \frac{k_w l}{420}\begin{bmatrix} 156 & 22l & 54 & -13l \\ 22l & 4l^2 & 13l & -3l^2 \\ 54 & 13l & 156 & -22l \\ -13l & -3l^2 & -22l & 4l^2 \end{bmatrix} \tag{5}$$

In the Pasternak model (**Figure 2**), the flexural behavior of this beam is given by Eq. (6) [5, 6]:

$$EI\frac{d^4w(x)}{dx^4} + p(x) - g\frac{d^2w(x)}{x^2} = q(x) \tag{6}$$

Solving Eq. (6) by FEM is expressed by Eq. (7):

$$([S_e] + [S_w] + [S_g])\{d\} = \{q\} \tag{7}$$

in which the expressions of the elastic stiffness matrix of the beam $[S_e]$ and subgrade reaction matrix $[S_w]$ are the same like those from Eqs. (4) and (5), and shear matrix $[S_g]$ is given by Eq. (8):

$$[S_g] = \frac{g}{30l}\begin{bmatrix} 36 & 3l & -36 & 3l \\ 3l & 4l^2 & -3l & -l^2 \\ -36 & -3l & 36 & -3l \\ 3l & -l^2 & -3l & 4l^2 \end{bmatrix} \tag{8}$$

Inserting the second parameter for the soil (shear stiffness) has the effect of increasing the stiffness of the beam (increasing the stiffness matrix terms). The stiffness matrices were obtained by considering a continuous bearing of the beam according to **Figure 3** and using a cubic displacement function [13, 14].

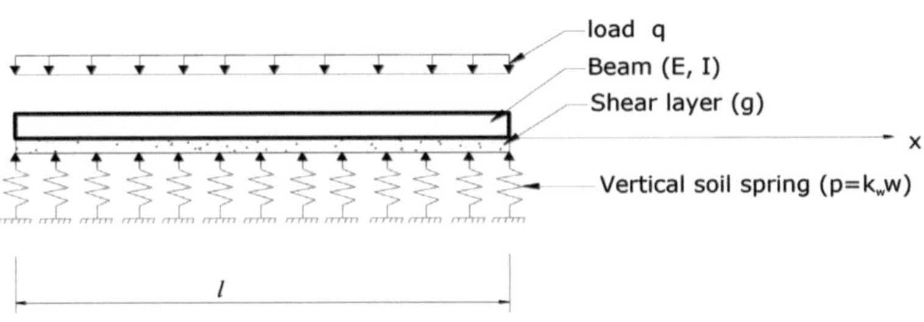

Figure 2.
The Pasternak model.

The meaning of the notations can be inferred from the following equation, which generally defines a stiffness matrix for a finite element with four degrees of freedom (9):

$$S = \begin{bmatrix} S_{11} & S_{12} & S_{13} & S_{14} \\ S_{21} & S_{22} & S_{23} & S_{24} \\ S_{31} & S_{32} & S_{33} & S_{34} \\ S_{41} & S_{42} & S_{43} & S_{44} \end{bmatrix} \tag{9}$$

2.2 Stiffness matrix calculation for support only at the nodes

In the calculation of the beams on elastic support by the finite element method, the determination of the stiffness matrix of the elastic foundation was determined by other authors [15] in the form (10):

$$[S_w] = \frac{k_w l}{2} \begin{bmatrix} 1 & 0 & 0 & 0 \\ 0 & 0 & 0 & 0 \\ 0 & 0 & 1 & 0 \\ 0 & 0 & 0 & 0 \end{bmatrix} \tag{10}$$

Eq. (10) is directly suggested by the calculation scheme of **Figure 4** where it can be seen that only the S_{11} and S_{33} elements of the stiffness matrix have values other than zero (a S_{ij} element of the stiffness matrix is the generalized force that develops in the direction i when a unit displacement or rotation is imposed in the direction j). Eq. (10) can also be obtained by solving Eq. (6) in which the shape functions have the expressions (11):

$$N_1(x) = \begin{matrix} 1, x < \frac{1}{2} \\ 0, \frac{1}{2} < x < 1 \end{matrix} \quad ; \qquad N_2(x) = 0, x \in [0, 1] \tag{11}$$

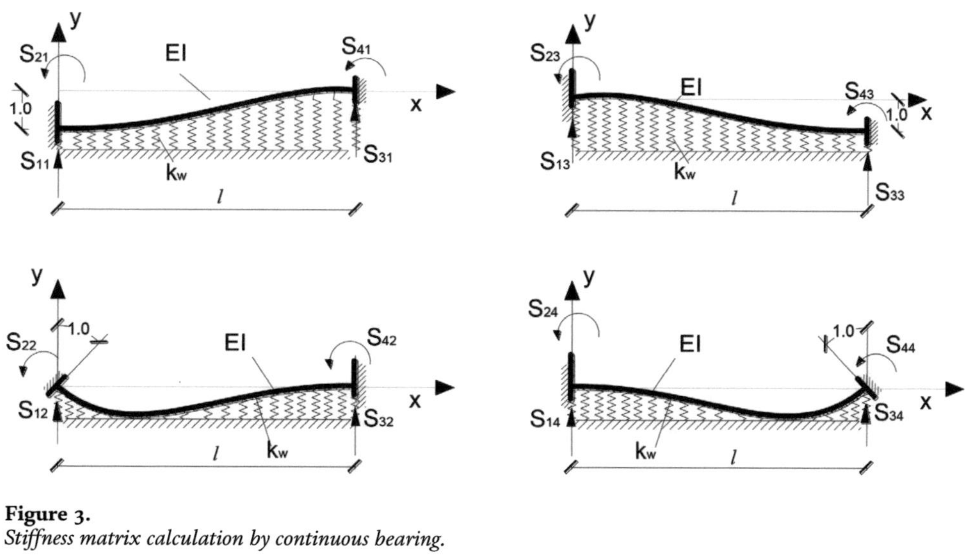

Figure 3.
Stiffness matrix calculation by continuous bearing.

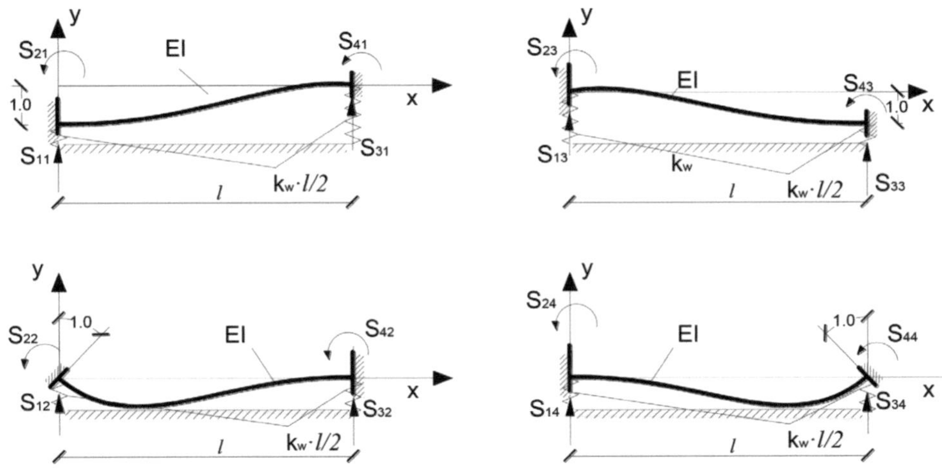

Figure 4.
Stiffness matrix calculation by nodes bearing.

$$N_3(x) = \begin{cases} 0, x < \frac{1}{2} \\ 1, \frac{1}{2} < x < 1 \end{cases}; \qquad N_4(x) = 0, x \in [0, 1]$$

With the same expressions (Eq. (10)) for shape functions shear matrix, $[S_g] = 0$. This result it can be understood that the absence of friction between springs considering that the springs are positioned only in the nodes (at a sufficient distance between them).

3. Conduit calculation as a beam

3.1 Calculation scheme

We have noticed that the influence of the earth parameters is quantified in the rigidity matrix of the FEM specific equations (1), (5), and (7). Depending on the multiparameter subgrade model, this influence is variable. Is it also significant for the behavior of the structure, if the values of the matrix elements, which quantify the rigidity of the earth, are comparable to those of the rigidity of the pipe? Further, for comparing the value of the matrix elements and their influence on the final result, we will compute these matrices for a real case [16]. Bottom discharge conduit from an embankment dam is considered as a structural element (Ibaneasa dam from Botosani county—Romania). The pipe is from reinforced concrete with polygonal section (inner rectangle, exterior trapeze) (see **Figure 5**).

The conduit was built in 9 m sections. We will make a calculation of a 9-m-long section in the central area of the dam. It is considered a single finite element between two longitudinal joints of length l (see **Figure 6**). The beam loading and support scheme are shown in **Figure 7**.

Numerical parameters are:

a. For the conduit parameters:

$A = 5.36\ m^2$, area of concrete section

$I_b = 6.67\ m^4$, moment of inertia

$E_b = 26$ GPa (for C12/15), modulus of elasticity

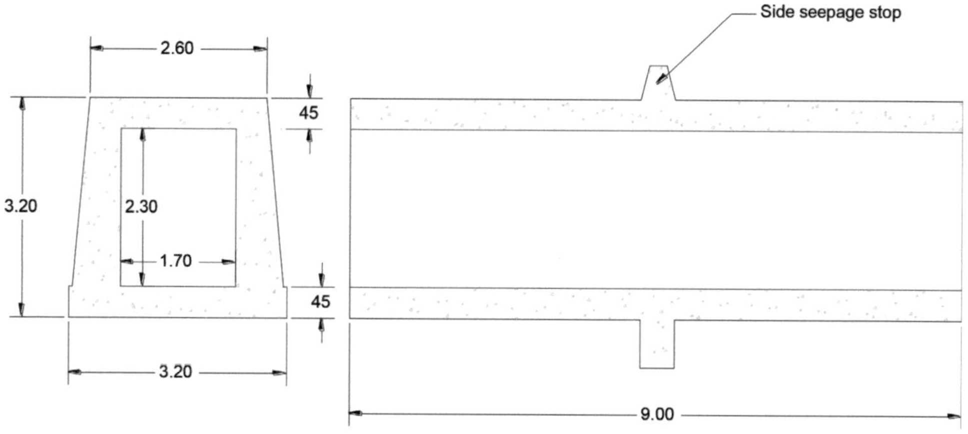

Figure 5.
Cross and longitudinal section by bottom discharge (reinforced concrete).

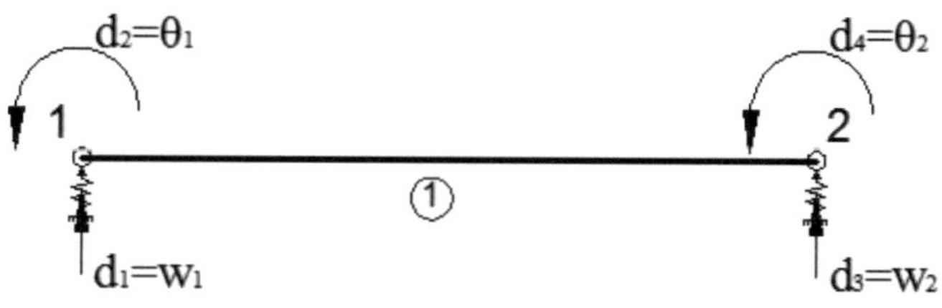

Figure 6.
Finite element scheme and degrees of freedom.

b. For the ground under conduit:

Each node will be thought of as a spring with its elasticity determined by:

$$k_s = B \cdot k \text{ in which} \tag{12}$$

B = 3.2 m is the width of the conduit

The marginal nodes will have the same coefficient of subgrade reaction as the other ones according to [15].

The coefficient of subgrade reaction according to Vesić apud Bowles [15] is given by:

$$\mathrm{k} = 0,65\sqrt[12]{\frac{\mathrm{E_p B^4}}{\mathrm{E_b I_b}}}\frac{\mathrm{E_p}}{\mathrm{B}\left(1-\mu_\mathrm{p}^2\right)} \tag{13}$$

Ground parameters are (silty clay): E_p = 35 MPa; μ_p = 0.35; γ_p = 19 kN/m^3.
With these values, it results in k_s = 3.2 · 5875 = 28.200 kN/m.
Shear modulus for shear layer in foundation is.

$$\mathrm{g} = \frac{\mathrm{E_p}}{2\left(1+\mu_\mathrm{p}\right)} = 13 \text{ MPa} \tag{14}$$

$$g_s = B \cdot g \tag{15}$$

Foundation parameters k and g may be calculated according [6, 7] with the following relations:

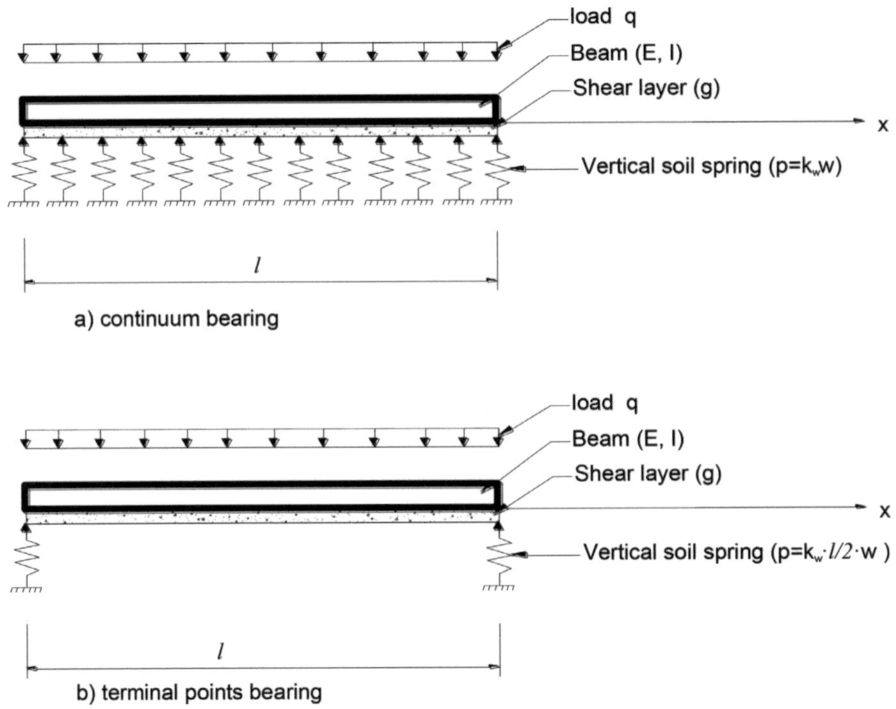

Figure 7.
Beam loading and support scheme.

$$k = \frac{E_P}{H} \tag{16}$$

$$g = \frac{E_P}{2\left(1 + \mu_p\right)} \frac{H}{2} \tag{17}$$

where H is depth to effective rigid base.

The effective rigid base is defined as the depth at which settlements caused by the structure can be taken to be zero. For decades it has been assumed that the "depth of influence" for settlement equivalent conceptually to the effective depth to rigid base is twice the width of a square loaded area and four times the width of an infinite strip [7].

With this assumptions H = 6.4 m; k = 5468 kN/m^2; g = 41.5 MPa.

The load on the conduit from the ground weight can be considered uniformly distributed. The load from the ground with its own pipe weight is 826 kN/m. With these parameters, it is necessary to determine the stresses in conduit schematized by the finite element from **Figures 6** and 7.

3.2 Solving the FEM equilibrium equation

The matrix equilibrium equation is (7)

$$\left([S_e] + [S_w] + [S_g]\right)\{d\} = \{q\} \tag{18}$$

which is written in form (1)

$$[S]\{D\} = \{Q\} \tag{19}$$

Solving the equation system (1), (7) is done by partitioning the S-matrices, D and Q separating the displacements according to the free degrees of freedom (2 and 4) by the degrees of freedom with elastic resistances (1 and 3) (see **Figure 6**):

$$\begin{bmatrix} S_{nn} & S_{nr} \\ S_{rn} & S_{rr} \end{bmatrix} \begin{Bmatrix} D_n \\ D_r \end{Bmatrix} = \begin{Bmatrix} Q_n \\ Q_r \end{Bmatrix} + \begin{Bmatrix} R_n \\ R_r \end{Bmatrix} \tag{20}$$

Eq. (20) results in two matrix equations representing the equations of the structure (21):

$$\begin{aligned} S_{nn}D_n + S_{nr}D_r &= Q_n + R_n \\ S_{rn}D_n + S_{rr}D_r &= Q_r + R_r \end{aligned} \tag{21}$$

where

$$D_n = \begin{Bmatrix} \theta_1 \\ \theta_2 \end{Bmatrix}; \quad D_r = \begin{Bmatrix} w_1 \\ w_2 \end{Bmatrix} \text{ are displacement subvectors} \tag{22}$$

$$Q_n = q \begin{Bmatrix} -\dfrac{l^2}{12} \\ \dfrac{l^2}{12} \end{Bmatrix} \qquad Q_r = q \begin{Bmatrix} -\dfrac{1}{2} \\ -\dfrac{1}{2} \end{Bmatrix} \text{ are load subvectors} \tag{23}$$

In solving Eqs. (21), there may be four situations depending on the connections of the structure with the soil [14]:

a. Fixed connections when $D_r = 0$ and from the first relationship in (21) result displacements and from the second relation (21) result the reactions.

b. In the case of known displacements of support, Dr has a known value from Eq. (21) that results displacements and reactions.

c. In the case of elastic connections at the nodes (see **Figures 6** and **7b**).

$$R_n = \begin{Bmatrix} R_2 \\ R_4 \end{Bmatrix} = \begin{Bmatrix} 0 \\ 0 \end{Bmatrix}; \quad R_r = \begin{Bmatrix} R_1 \\ R_3 \end{Bmatrix} = k_s \begin{Bmatrix} d_1 \\ d_3 \end{Bmatrix} = k_s D_r \text{ are reactions.} \tag{24}$$

Eq. (24) can be written as

$$R_r = \begin{Bmatrix} R_1 \\ R_3 \end{Bmatrix} = \begin{bmatrix} k_s & 0 \\ 0 & k_s \end{bmatrix} \begin{Bmatrix} d_1 \\ d_3 \end{Bmatrix} = [k_s] D_r \tag{25}$$

By replacing (24) in (21), we get

$$\begin{aligned} S_{nn}D_n + S_{nr}\frac{1}{k_s}R_r &= Q_n \\ S_{rn}D_n + \left(S_{rr}\frac{1}{k_s} - I\right)R_r &= Q_r \end{aligned} \tag{26}$$

From the last Eq. (26) results

$$R_r = \left(\frac{1}{k_s} S_{rr} - I\right)^{-1} (Q_r - S_{rn} D_n) \tag{27}$$

which introduced in the first Eq. (26) leading to the calculation of the displacements, which is depicted below

$$S_{nn} D_n + S_{nr} \frac{1}{k_s} \left(\frac{1}{k_s} S_{rr} - I\right)^{-1} (Q_r - S_{rn} D_n) = Q_n$$

$$S_{nn} D_n + S_{nr} \frac{1}{k_s} \left(\frac{1}{k_s} S_{rr} - I\right)^{-1} Q_r - S_{nr} \frac{1}{k_s} \left(\frac{1}{k_s} S_{rr} - I\right)^{-1} S_{rn} D_n = Q_n$$

$$\left[S_{nn} - S_{nr} \frac{1}{k_s} \left(\frac{1}{k_s} S_{rr} - I\right)^{-1} S_{rn}\right] D_n = Q_n + S_{nr} \frac{1}{k_s} \left(\frac{1}{k_s} S_{rr} - I\right)^{-1} Q_r$$

Displacements in the directions of unrestrained degrees of freedom are

$$D_n = \left(S^*_{nn}\right)^{-1} Q^*_n \tag{28}$$

where

$$S^*_{nn} = S_{nn} - S_{nr} \frac{1}{k_s} \left(\frac{1}{k_s} S_{rr} - I\right)^{-1} S_{rn}; \qquad Q^*_n = Q_n + S_{nr} \frac{1}{k_s} \left(\frac{1}{k_s} S_{rr} - I\right)^{-1} Q_r \tag{29}$$

d. In case of elastic connection and continuous support

In this case, (**Figure 7a**, continuum bearing) the rotations are not independent movements; they depend on the rotational stiffness of the ground. In the situation of (c), the rigidity of the displacement matrix had the form

$$[k_s] = \begin{bmatrix} k_s & 0 \\ 0 & k_s \end{bmatrix} \tag{30}$$

In case (d), we will obtain the matrix $[k_s]$ by static condensation of the sum of matrices:

$$[K] = [S_w] + [S_g] \tag{31}$$

For condensation we will use the Guyan method [13]. The condensed matrix K_r is obtained from the matrix K whose terms have the meaning below. DOFs (degrees of freedom) retained are 1 and 3, and DOFs omitted are 2 and 4 (rotation):

$$K = \begin{bmatrix} K_{11} & K_{12} & K_{13} & K_{14} \\ K_{21} & K_{22} & K_{23} & K_{24} \\ K_{31} & K_{32} & K_{33} & K_{34} \\ K_{41} & K_{42} & K_{43} & K_{44} \end{bmatrix} \tag{32}$$

Calculation relation for condensed rigidity matrix of earth parameters is (33) [13].

$$K_r = K_{rr} - K_{ro} K_{oo}^{-1} K_{ro}^T \tag{33}$$

where

$$K_{rr} = \begin{bmatrix} K_{11} & K_{13} \\ K_{31} & K_{33} \end{bmatrix} \quad K_{rn} = \begin{bmatrix} K_{12} & K_{14} \\ K_{32} & K_{34} \end{bmatrix} \quad K_{oo} = \begin{bmatrix} K_{22} & K_{24} \\ K_{42} & K_{44} \end{bmatrix} \tag{34}$$

The condensed rigidity matrix includes the effect of the rigidity of the foundation on the rotation of the beam ends.

3.3 Obtained results

3.3.1 Results for the Winkler (single-parameter) scheme

Stiffness matrix is

$$[S] = [S_e] + [S_w]$$

$$S_e = \begin{pmatrix} 2.855\times10^6 & 1.285\times10^7 & -2.855\times10^6 & 1.285\times10^7 \\ 1.285\times10^7 & 7.708\times10^7 & -1.285\times10^7 & 3.854\times10^7 \\ -2.855\times10^6 & -1.285\times10^7 & 2.855\times10^6 & -1.285\times10^7 \\ 1.285\times10^7 & 3.854\times10^7 & -1.285\times10^7 & 7.708\times10^7 \end{pmatrix}$$

$$S_w = \begin{pmatrix} 6.285\times10^4 & 7.977\times10^4 & 2.175\times10^4 & -4.713\times10^4 \\ 7.977\times10^4 & 1.305\times10^5 & 4.713\times10^4 & -9.789\times10^4 \\ 2.175\times10^4 & 4.713\times10^4 & 6.285\times10^4 & -7.977\times10^4 \\ -4.713\times10^4 & -9.789\times10^4 & -7.977\times10^4 & 1.305\times10^5 \end{pmatrix}$$

$$S = \begin{pmatrix} 2.917\times10^6 & 1.293\times10^7 & -2.833\times10^6 & 1.28\times10^7 \\ 1.293\times10^7 & 7.721\times10^7 & -1.28\times10^7 & 3.844\times10^7 \\ -2.833\times10^6 & -1.28\times10^7 & 2.917\times10^6 & -1.293\times10^7 \\ 1.28\times10^7 & 3.844\times10^7 & -1.293\times10^7 & 7.721\times10^7 \end{pmatrix}$$

The displacements of the conduit ends are:

$$D := \begin{pmatrix} -0.068 \\ -3.78\times10^{-4} \\ -0.068 \\ 3.78\times10^{-4} \end{pmatrix} = \begin{Bmatrix} w_1 \\ \theta_1 \\ w_2 \\ \theta_2 \end{Bmatrix}$$

Displacements w_1 and w_2 are measured in [m] and rotation in [rad]
The rigidity matrix (Eq. (30)) has the value

$$k_S = \begin{pmatrix} 3.008 \times 10^4 & 0 \\ 0 & 3.008 \times 10^4 \end{pmatrix}$$

The displacement of the center of the beam (conduit) is

$$w = -0.0689 \ [m]$$

Displacement in the center of the beam is calculated using shape functions according to relation:

$$w(l/2) = [N_1 \ N_2 \ N_3 \ N_4]\{w_1 \quad \theta_1 \quad w_2 \quad \theta_2\}^T \tag{35}$$

3.3.2 Results for multiple-parameter subgrade model scheme (Pasternak)

The stiffness matrix is

$$[S] = [S_e] + [S_w] + [S_g] \tag{36}$$

The stiffness matrix values are:

$$S_e = \begin{pmatrix} 2.855 \times 10^6 & 1.285 \times 10^7 & -2.855 \times 10^6 & 1.285 \times 10^7 \\ 1.285 \times 10^7 & 7.708 \times 10^7 & -1.285 \times 10^7 & 3.854 \times 10^7 \\ -2.855 \times 10^6 & -1.285 \times 10^7 & 2.855 \times 10^6 & -1.285 \times 10^7 \\ 1.285 \times 10^7 & 3.854 \times 10^7 & -1.285 \times 10^7 & 7.708 \times 10^7 \end{pmatrix}$$

$$S_W = \begin{pmatrix} 6.285 \times 10^4 & 7.977 \times 10^4 & 2.175 \times 10^4 & -4.713 \times 10^4 \\ 7.977 \times 10^4 & 1.305 \times 10^5 & 4.713 \times 10^4 & -9.789 \times 10^4 \\ 2.175 \times 10^4 & 4.713 \times 10^4 & 6.285 \times 10^4 & -7.977 \times 10^4 \\ -4.713 \times 10^4 & -9.789 \times 10^4 & -7.977 \times 10^4 & 1.305 \times 10^5 \end{pmatrix}$$

$$S_g = \begin{pmatrix} 5.531 \times 10^3 & 4.148 \times 10^3 & -5.531 \times 10^3 & 4.148 \times 10^3 \\ 4.148 \times 10^3 & 4.978 \times 10^4 & -4.148 \times 10^3 & -1.244 \times 10^4 \\ -5.531 \times 10^3 & -4.148 \times 10^3 & 5.531 \times 10^3 & -4.148 \times 10^3 \\ 4.148 \times 10^3 & -1.244 \times 10^4 & -4.148 \times 10^3 & 4.978 \times 10^4 \end{pmatrix}$$

$$S = \begin{pmatrix} 2.923 \times 10^{6} & 1.293 \times 10^{7} & -2.838 \times 10^{6} & 1.28 \times 10^{7} \\ 1.293 \times 10^{7} & 7.726 \times 10^{7} & -1.28 \times 10^{7} & 3.843 \times 10^{7} \\ -2.838 \times 10^{6} & -1.28 \times 10^{7} & 2.923 \times 10^{6} & -1.293 \times 10^{7} \\ 1.28 \times 10^{7} & 3.843 \times 10^{7} & -1.293 \times 10^{7} & 7.726 \times 10^{7} \end{pmatrix}$$

The condensed rigidity matrix of earth parameters, according relation (33) is:

$$k_S = \begin{pmatrix} 2.871 \times 10^{4} & 472.813 \\ 472.813 & 2.871 \times 10^{4} \end{pmatrix}$$

The displacements of the conduit ends are:

$$D := \begin{pmatrix} -0.0657 \\ -3.694 \times 10^{-4} \\ -0.0657 \\ 3.694 \times 10^{-4} \end{pmatrix} = \begin{Bmatrix} w_1 \\ \theta_1 \\ w_2 \\ \theta_2 \end{Bmatrix}$$

Displacements w_1 and w_2 are measured in [m] and rotation in [rad]. The displacement of the center of the beam (conduit) is

$$w = -0.0665 \quad [\mathrm{m}]$$

3.4 Comments about conduit calculation as a beam

The calculation of rigidity for the conduit and for the terrain is common. Solving equilibrium Eq. (7) for situations (a), (b), and (c) is usual. For continuous support (d) we used condensation of stiffness matrices, the method that I consider new, through which includes the rigidity of rotation of the bar and the rigidity of the terrain in the stiffness of the bar ends. In this way we have reduced the number of unknowns—reactions, and solving the equation in the case (d) becomes similar to the solution in the case (c).

The purpose of calculating rigidity matrices and conduit (beam) displacements, for the two methods of determining the ground parameters (the Winkler and multiple-parameter subgrade model, the Pasternak), is to highlight the small difference between the calculated values. This small difference between displacements (4%) is due to the high rigidity (high moment of inertia) of the conduit.

A first conclusion is that for structures with a high transversal moment of inertia, the method of determining the parameters of the earth is not very important. The difference between the elements of rigidity matrices for the structure and for the Winkler earth is about 100 times and between the Winkler earth and the Pasternak rigidity is about 10 times.

A second conclusion is that the use of the SAP 2000 finite element software which includes the Winkler spring model is acceptable for bottom discharge conduit due to their high rigidity (of the conduit).

From these considerations, we can identify two calculation methods for this type of conduits. In a first method, the pipeline is calculated as a beam supported by the Winkler-type springs. You can use any finite element software that allows you to use springs. The joints between the conduit segments are introduced into the calculation scheme with their specific conditioning (restraints). In a second method the calculation can be done with a geotechnical software (e.g., GeoStudio) in which the modulus of elasticity is introduced as input parameters and the calculation is performed in a plane strain condition [17]. In this calculation method, somehow the influence of the joints between the pipe sections must be simulated. The following relationship is proposed for calculating the equivalent inertia moment of the pipe in which joints are taken into account:

$$I_e = I\frac{1}{2n} \tag{37}$$

where I_e is the equivalent inertial momentum; I is the current section inertial momentum; n is the number of joints.

4. Conduit calculation example

The calculation will be made for the bottom discharge conduit of the Tungujei earth dam in Iasi County, Romania, designed as a cross-sectional structure of three rectangular reinforced concrete cassettes (**Figure 8**) [18]. In the longitudinal profile, the structure has joints between 5 and 10 m. The joints are sealed with PVC tape, and the reinforcement has no continuity in the joint.

The calculation scheme is the Winkler elastic beam and the EngiLab Beam 2D software [19]. The conduit consists of 2 reinforced concrete sections of 5 m and another 11 sections of 10 m length each (**Figure 9**). Between the sections there were joints of 2.5 cm, and the continuity is achieved with the sealing tape.

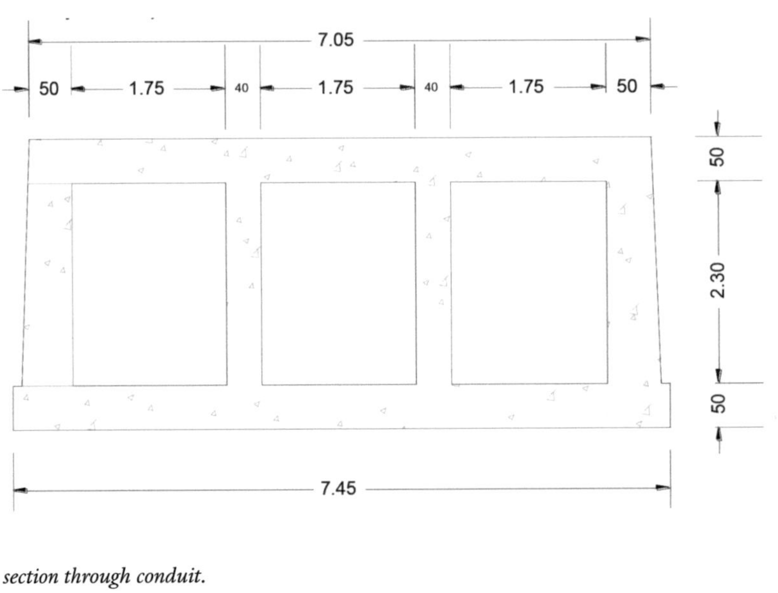

Figure 8.
Transverse section through conduit.

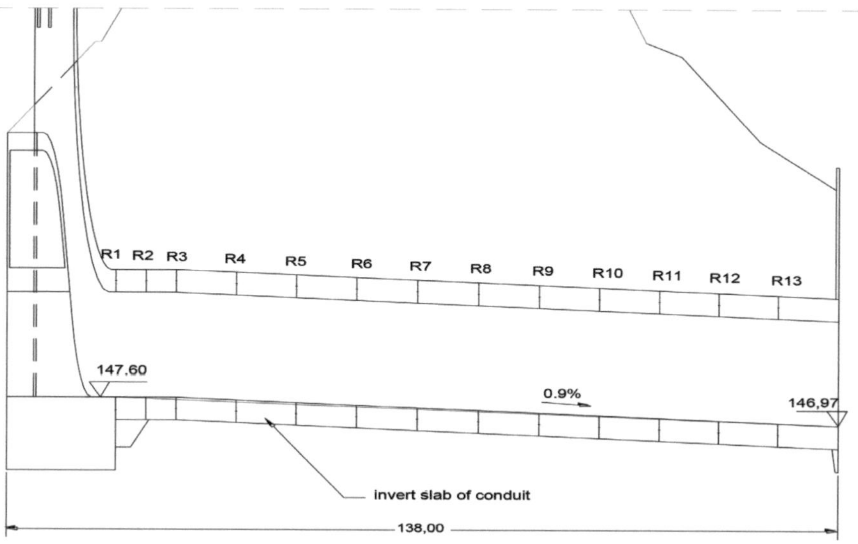

Figure 9.
Longitudinal section through conduit.

The vertical load on the pipe is after Marston [4]:

$$P_v = C_e \gamma D_e^2, \text{ where:} \tag{38}$$

$$C_e = \frac{e^{2K\mu(H/D_e)} - 1}{2K\mu} \quad \text{or} \tag{39}$$

$$C_e = \frac{e^{2K\mu(H_e/D_e)} - 1}{2K\mu} + \left(\frac{H}{D_e} - \frac{H_e}{D_e}\right) e^{2K\mu(H_e/D_e)} \tag{40}$$

where $K = tg^2(45^o - j/2)$ is Rankine's lateral pressure coefficient; $\mu = tg\,\varphi$ is the coefficient of friction of the earth; H_e is the position of the plan of equal settlement.

Marston determined the existence of a horizontal plane above the conduit where the shearing forces are zero. This plane is called the *plane of equal settlement*. Above this plane, the interior and exterior prisms of soil settle equally.

D_e is the outer diameter (width) of the pipe.

The relation (39) is valid for He > H (the plane of equal compression is imaginary) and the relation (40) is valid for He <H (the additional load relative to the weight of the earth column) depend on the top-down friction forces appearing in the earth column above the.

conduit in the vertical planes tangent to the pipe. In the case of the Tungujei dam, the thickness of the filling of 15 m over a 7.45 m width of the earth column is per linear meter of the pipe. According to the Marston relationship, it follows (considering the plane of equal compaction to the surface $H = H_e$).

$$Ce = 2.64; Pv = 2784\ kN/m$$

The pipeline calculation parameters are
$A = 11.67\ m^2$, area of the concrete conduit
$I_b = 16.05\ m^4$, moment of inertia for section
$E_b = 26$ GPa (for concrete C12/15), modulus of elasticity for concrete
Foundation parameters *k* may be calculated according Horvath [6] with the following relations:

$$k = \frac{E_p}{H} \tag{41}$$

where E_p = 18,000 kN/m^2 earth (foundation) modulus; H = 2 × 7.5 m = 15 m, depth of influence; k = 1200 kN/m^3.

When loading with the weight of the ground, the weight of the conduit cassette was added.

Two pipeline calculation schemes were used between the R3 and R12 joints (see **Figure 9**). The 5 m sections were removed from the calculation scheme and the terminal sections embedded in the portal.

In the first diagram, the nine sections of the pipe were considered to be articulated at the ends. A maximum of 32.9 cm settlement resulted (**Figures 10–12**).

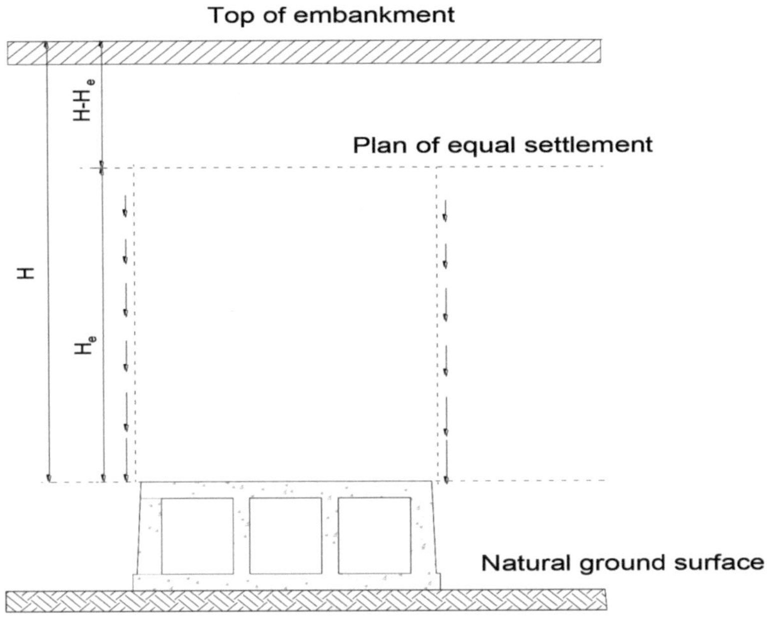

Figure 10.
Vertical loads that the earth filling exerts on the conduit.

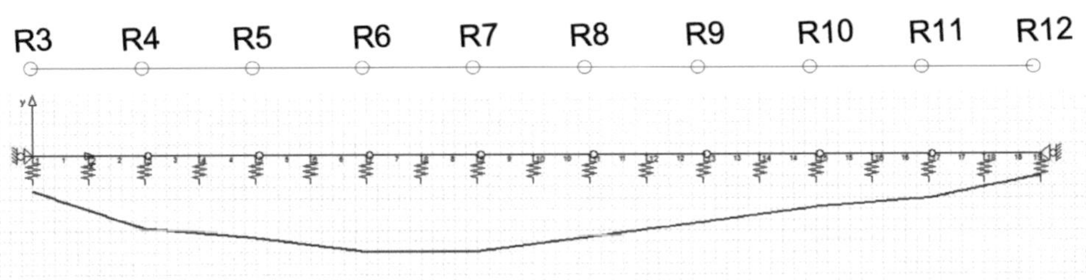

Figure 11.
Conduit displacement with articulations in joints.

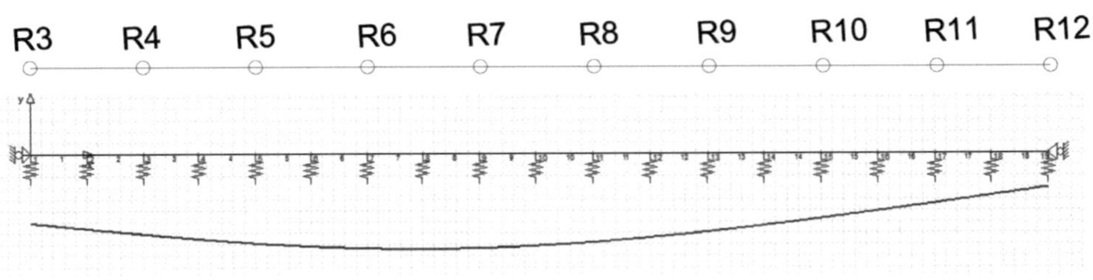

Figure 12.
Conduit displacements without articulations and equivalent moment of inertia.

In the second calculation scheme, the nine continuous pipe sections were considered without interruptions in joints. A maximum of 32.6 cm settlement resulted.

The equivalent moment of inertia is calculated with Eq. (37):

$$I_e = I\frac{1}{2n} = 16.04\frac{1}{2 \times 8} = 1m^4$$

5. Conclusions

We calculated conduits through FEM using analytical calculation for a finite element and calculation with a software (EngiLab Beam). In the calculation for a finite element, we have highlighted the stiffness matrices to understand the large value difference between the terms of the matrices and the physical significance of this difference.

The low rigidity of the ground slightly influences the state of stresses and deformations in the conduit. It follows that it is sufficient that the parameters of the earth are defined by the Winkler-type springs.

A first conclusion is that computational programs that allow the use of the Winkler-type springs give results with good accuracy (e.g., below 5%) for this type of structure. The overall view at this point is that the Winkler model is outdated.

The difference in conduit displacements calculated with the two methods of estimating the ground characteristics (the Winkler and multiple-parameter subgrade model) is small. It means that the shear stiffness introduced by multiple-parameter subgrade models does not significantly change the result. It is therefore reasonable to use the Winkler parameters for the terrain characteristics. These considerations apply to conduits where the stiffness of the conduit (measured in the stiffness matrix) is 100 times higher than the rigidity of the earth (also expressed by the stiffness matrix).

The calculation of bottom discharge conduit as well as other long hydrotechnical structures is done with reasonable accuracy in the plane strain state. Finite element programs calculate stress state and deformations with this scheme. In the case of conduits, the influence of the joints between the sections must be modeled. If this is not done, the results are not credible. In the calculated example, we used two schemes: one with the pipe provided with joints and with the respective struts and the second with the pipe without joints (continuous) and with the inertia moment adjusted with the relation (37). The difference between the results obtained with these two schemes is small (1%). Models of finite element type are credible if the conduit is made jointless, which is unlikely.

Resuming the findings of this work are:

Using the rigidity matrix condensation, to solve the equilibrium equation for beams with deformable supports, facilitates solving the equation system (21).

We appreciate that the use of the Winkler-type springs is suitable for long rigid structures.

Calculation of conduit sections in plane strain state leads to credible results only if the moment of inertia is adjusted in accordance with pipe joints. For this we propose a relation of calculating the moment of equivalent inertia (37). If the inertia moment is not adjusted, the result obtained is not credible.

Conflict of interest

The author declares that there is no conflict of interest.

Author details

Costel Boariu
Faculty of Hydrotechnical Engineering, Geodesy and Environmental Engineering, Gheorghe Asachi Technical University of Iasi, Romania

*Address all correspondence to: costelboariu@gmail.com

References

[1] Popovici A. Dams for Water Storage. (in Romanian). Vol. 2. Bucuresti: Editura Tehnica; 2002

[2] Federal Emergency Management Agency. Conduits through Embankment Dams, Best Practices for Design, Construction, Problem Identification and Evaluation, Inspection, Maintenance, Renovation, and Repair, Technical Manual. Denver Colorado; 2005

[3] U.S. Army Corps of Engineers, Culverts, Conduits, and Pipes, EM 1110–2-2902, 1997, 1998

[4] Moser AP, Folkman SL. Buried Pipe Design. 3rd ed. New York: Mc Graw Hill; 2008. DOI: 10.1036/007147689X

[5] Teodoru IB, Musat V. Numerical Modeling of Structure Soil Interaction. Foundation Beams (in Romanian). Romania: Editura Politehnium Iasi; 2009

[6] Horvath JS. Soil-Structure Interaction Research Project [Internet]. 2002. Available from: www.engineering.manhattan.edu/civil/CGT.html [Accessed: 12 September 2015]

[7] Horvath JS, Colasanti RJ. Practical Subgrade Model for Improved Soil-Structure Interaction Analysis: Model Development. International Journal of Geomechanics. 2011;**11**(1):59-64

[8] Bureau of Reclamation, Embankment Dams—Embankment Design, Design Standards No. 13, Chapter 2, 2012

[9] Bureau of Reclamation. Appurtenant Structures for Dams (Spillways and Outlet Works) Design Standards, October 2011

[10] Bureau of Reclamation. Design of Small Dams, 3rd ed. Washington DC; 1987

[11] CSI. SAP2000 Analysis Reference Manual. Berkeley, CA: Computers and Structures, Inc.; 2016

[12] Geo-Studio. Stress-Deformation Modeling with SIGMA/W. Canada: Calgary, Alberta; 2012

[13] Chandrupatla TR, Belegundu AD. Finite Elements in Engineering. 3rd ed. Prentice Hall; 2002

[14] Jerca SH, Ungureanu N, Diaconu D. Numerical Methods in Construction Design (in Romanian). Romania: UT Gh. Asachi Iasi; 1997

[15] Bowles JE. Foundation Analysis and Design. 5th ed. Singapore: McGraw-Hill Book Co; 1996

[16] Boariu C. Soil structure interaction calculus, for rigid hydraulic structures, using FEM. IOSR Journal of Mechanical and Civil Engineering (IOSR-JMCE). 2015;**12**:60-68

[17] Kasey J. Stress Analysis of Soils Adjacent to Selected Outlet Conduit Configurations for Small Embankment Dams [Internet]. 2015. Available from: http://dnrc.mt.gov/divisions/water/operations/dam-safety/OutletResearchFinalReport2015.08.18.pdf [Accessed: 07 July 2018]

[18] Boariu C, Bofu C. Safety Assurance on Existing Dams. Case Study—Tungujei Dam. In: 16th National Technical-Scientific Conference on Modern Technologies for the 3rd Millennium. ISBN: 978-88-87729-41-2; Oradea, Romania; Date: MAR 23–24, 2017. pp. 7-12

[19] EngiLab. EngiLab Beam.2D 2015 (v2.2) software. 2015. http://www.engilab.com/

Chapter 4

Review of Methods of Measuring Streamflow Using Hydraulic Structures

Adeniyi Ganiyu Adeogun and
Apalando Abdulrasaq Mohammed

Abstract

The measurement of streamflow is very critical to hydraulic engineers and hydrologists as it provides vital information for environmental monitoring issues connected to water resources. The objective of this study is to examine various means of measuring streamflow specifically application of hydraulic structures installed across the direction of flow. Weirs are restricted to small rivers where the provision for sufficient head and constriction in the river is acceptable. Sharp-crested weir is easy to construct, and it is commonly used as a flow measuring device in an open channel. Flumes are self-cleaning due to the fact that the flow velocity through a flume is usually high. Traditionally, flume is used in measuring flow in agricultural systems, and it requires low maintenance cost. It has capacity to measure more flow rates than weir. Accurate streamflow measurement using flume is within ±2–5% while that of weir is ±2%. Generally, flumes are employed to determine discharge where weirs are not useful.

Keywords: flumes, notch, orifices, streamflow, weirs

1. Introduction

Streamflow is very important in estimating hydrology cycle [1]. In practice, hydraulic structures are installed in open channels or rivers with a free water level to estimate discharge based on the measured upstream water level [2]. The main critical factors in constructing hydraulic structures across an open channel throughout the world are need for the reliable source of water supply, flood control, irrigation schemes, recreation activities and hydropower generation [3]. Technological interventions are needed for harnessing, conserving and proper management of water resources. Application of hydraulics structures in measuring streamflows in open channels is very important. Flumes and weirs are used in measuring streamflow in natural and artificial channels. Streamflow is manually or automatically measured.

Flumes and weirs are designed and constructed to change flow regime from one state to another. The parameters of flow measured in laboratory experiment of flume and models of weirs or spillways can be applied to an open channel prototype in real-life situation by manipulating the various scale models in the laboratory experiment [4].

Spillways are designed to safely convey flood to water course downstream from the dams and to prevent overtopping of the dams. The selection, design and construction of a particular type of spillway is carried on specific purpose of the project, hydrology of the area, topography, geologic conditions, dam safety and project economy. Provisions of hydraulically efficient and structurally strong spillway are vital for dam safety. In the olden days, many dams were operated with little hydrological information. Significant improvements made in meteorology and hydrology have updated control of flooding. The impact of this backwardness made many dams to have inadequate spillway capacities [5].

2. Methodology

The approach adopted in this work was reviewing and examining various hydraulic structures used in measuring streamflow in open channel structures and suggesting the most efficient one.

2.1 Streamflow measurement

Measurement of streamflow is essential in river basin planning, management and pollution mitigation. Several methods have been used for measuring streamflow. One of the methods used is the velocity area method which involves division of river into a number of segments. Flow in each segment is determined with the product of the area of the segment and the average velocity of stream at that location. The total discharge is then estimated by addition of the outcome for each segment of the river. The average velocity is measured using appropriate equipment such as water current metre. The corresponding area is computed through measurement of distances from a reference point on the riverbank.

Alternatively, streamflow is determined using hydraulic structures constructed across the river flow, and it requires establishing a good association between the head and the flow using empirical means as carried out in the laboratory. The two methods have certain limitations and are not applicable in all the circumstances. For example, velocity area method requires condition, which produces a stable depth discharge relationship, and the use of flume and weir is restricted to small rivers where there the provision for sufficient head and constriction in the river is acceptable.

Determination of flow in an experimental open channel was done by [6] using mean velocity equation. It was discovered that the proposed method was more accurate in estimating discharge, when compared with the conventional formulae.

2.1.1 Direct methods of streamflow measurement

Direct measurement of flow is achieved through several approaches; however, the use of any of such methods is based on some factors like size of the stream and the availability of equipment and expertise [7]. Generally, the section at which the discharge will be measured would be carefully selected so that the river reach is straight and is free of large obstacles that can impede the streamflow. Also, areas around or immediately downstream of existing hydraulic structures should be avoided so that the changes in flow conditions around these areas will not affect the streamflow measurement.

2.1.2 Volumetric method

Volumetric method is applied to measure small quantity of streamflow in ditches. In this method, the time taken to fill a containing vessel of known volume is measured, and the most accurate method of discharge measurement is to simply measure the time required to fill a container of known volume. This measurement can be taken repeatedly and the average streamflow can be estimated from the data [8].

2.1.3 Dilution method

In this method, a concentrated tracer solution like salt or dye of specific concentration is added to the river. This is followed by chemical analysis to determine its dilution after it has mixed completely with the stream and produced a uniform final output in the stream. The selection of the tracer to be used is based on meeting certain criteria. It should be easily detected and measured accurately. It should also be conservative. Salt and dye have exhibited these properties and have been in use for many years ago. Irrespective of the tracer selected, it is necessary to get required permission from relevant agency of government before addition of such tracer to body of water used for municipal water supply [8]. The expression used to determine the discharge using the dilution method is presented in Eq. (1):

$$Q = q\frac{(C_1 - C_2)}{(C_2 - C_o)} \tag{1}$$

where Q is the stream discharge (m^3/s), q is the tracer injection rate, C_1 is the tracer concentration in injection, C_2 is the final concentration of tracer in the stream and C_o is the background tracer concentration in the stream.

2.2 Indirect methods of streamflow measurement

These methods involve using various empirical formulae when it is impossible to measure discharge. Empirical formulae such as Chezy, Manning, and Strickler formulae are commonly used.

2.2.1 Chezy equation

One of the earlier formulas to evaluate the flow of water in river is Chezy equation. The formula was proposed in 1768 by a French engineer when designing a water supply canal in Paris [8]. Chezy coefficient depends on Reynolds number and boundary roughness. The Chezy expression for determining discharge in an open channel is given in Eq. (2):

$$Q = AC\sqrt{R\ S_{\circ}} \tag{2}$$

where Q is the stream discharge (m^3/s), A is the channel cross-sectional area (m^2), C is the Chezy coefficient (dimensionless), R is the hydraulic radius (m) and S_o is the channel bed slope (dimensionless).

2.2.2 Manning equation

The formula is named after an Irish engineer, Robert Manning, in 1889. The Manning equation is commonly used for the design of ditches carrying water. The Manning expression for calculating stream discharge in an open channel is presented in Eq. (3):

$$Q = \frac{1}{n} A R^{2/3} S_o^{1/2} \tag{3}$$

where Q is the stream discharge (m^3/s), A is the cross-sectional area of the channel (m^2), R is the hydraulic radius (m), S_o is the slope of the channel bed and n is the Manning roughness coefficient of the channel.

3. Flow measurement using hydraulic structures

Flow measurement using hydraulic structures involves the placement of a selected hydraulic structure such as weirs or flumes across a river channel. This is to generate flow properties that can be used to develop relationships between flow rate and water levels at certain section along the stream. This method uses a hydraulic structure constructed across a channel to produce flow properties that are characterized by relationships between the water level measurement at some location and the flow rate of the stream. The streamflow is estimated by taking measurement of the water surface level in or near the restriction of the hydraulic structure.

3.1 Weirs

Weirs are hydraulic structures which water flows over. They have advantage of being relatively lower in cost and relatively simple to construct. Weirs can be easily installed in open channels and a level of accuracy can be achieved when used appropriately. However, it is normally operated with a significant head loss, and its degree of accuracy can be affected by variation in approach velocity of water in flow channel [4]. A weir must be periodically cleaned to prevent sediment deposits at the upstream side of the weir, which will have adverse effect on the weir accuracy. Errors resulting from the approximations of discharge are corrected by means of a coefficient of discharge. Through experiment, the coefficient of discharge of a weir has been found to vary with the approach head; the results of the extensive tests with a V-notch weir produce results, which showed the variations [7].

3.1.1 Broad-crested weirs

Broad-crested weir is also known as long base weir. It is made up of an obstruction in the form of a raised portion of the bed, and it spans the full width of the channel with a crest sufficiently broad in the direction of flow for the surface of the liquid to become parallel to the crest of the weir [8, 9]. Broad-crested weirs are very robust structures and are generally constructed using reinforced concrete. The flow upstream is tranquil and conditions downstream allow a free fall over the weir. The flow characteristics of rectangular broad-crested weirs with sloped upstream face were studied by [10]. The results showed that decreasing upstream slopes from 90 to 10° resulted in increasing discharge coefficient values and dissipation of the separation zone. Flow was simulated over broad-crested weirs using 2D and 3D computational fluid dynamic models. Results revealed that the 3D required more computation period than 2D [11].

Investigation on the effects of surface roughness sizes on the discharge coefficient for broad-crested weirs was carried out by [12]. Results showed that the logical negative effect of roughness increased with discharge for different lengths of the weir. The flow over a broad-crested weir in subcritical flow conditions was studied [13]. It was found that the discharge coefficient of a rectangular broad-crested weir was related to upstream total head above the crest, length of weir and channel breadth. Also, [14] studied the discharge relations for rectangular broad-crested weirs. Results showed that a discontinuity occurred in head-discharge ratings because the section width experienced a break in slope when the flow entered the outer section. Values of coefficient of discharge (C_d) obtained from the experiments on compound broad-crested weirs are lower than those of a broad-crested weir with a rectangular cross section. **Figure 1** illustrates flow pattern over a broad-crested weir. In broad-crested weir, the flow over the crest of the weir is critical so the discharge equation is not the same as that of the sharp-crested weir. The modified equation of the broad-crested weir is presented in Eq. (4). Rectangular broad-crested weirs are commonly employed to measure discharge in irrigation canals especially in developing countries [15]:

$$Q = C B \sqrt{g} \left(\frac{2}{3} H\right)^{3/2} \tag{4}$$

where Q is the discharge over the weir (m^3/s), B is the width of the weir (m), C is the discharge coefficient (dimensionless), H is the head of water over the weir model, upstream (m) and g is the acceleration of gravity = 9.81 m/s^2.

3.1.2 Sharp-crested weir

Sharp-crested weir is regarded as the simplest form of flow measuring device over spillway in the measurement of flow in open channel. The characteristic of flow over the sharp-crested weir was recognised early in hydraulics engineering as the basis for design of round-crested overflow spillway. The shape of flow (nappe) over the sharp-crested weir can be represented by the principle of projectile. A sharp-crested weir is simple to instal and frequently used as a flow measuring device in an open channel. The determination of the discharge coefficient over sharp-crested weir was conducted by [16]. The results revealed that on the average, the coefficient of discharge was 0.7. Also, [17] determined the discharge coefficient

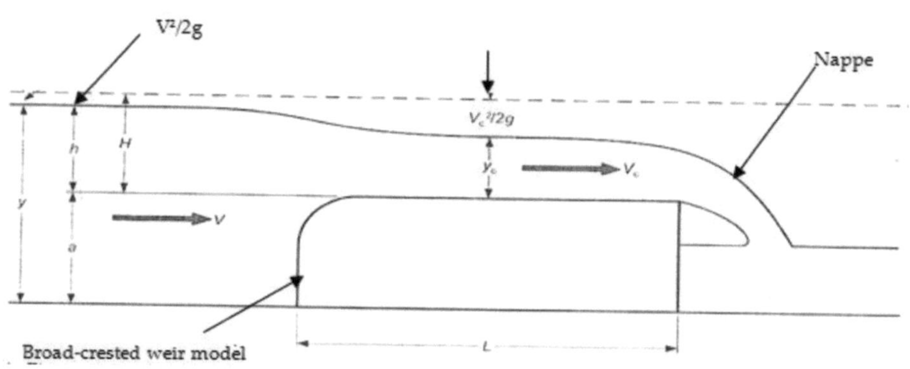

Figure 1.
Flow pattern over a broad-crested weir.

in an inclined rectangular sharp-crested weir using experimental and numerical simulation. Results revealed that the discharge coefficient of the weir increases with the increase in inclination of the weir plane. The discharge coefficient for a sharp-crested weir was investigated to vary between 0.61 and 0.73 [18].

A commonly used sharp-crested weir structure for measuring streamflow in irrigation and drainage channels is rectangular in shape [19]. In the sharp-crested experiment, the relationship between water level over the weir crest and discharge can be established; also the coefficient of the discharge for the weir can be determined. At the upstream of the sharp-crested weir, the velocity of all moving water elements is nearly uniform and parallel to the channel bed [19]. **Figure 2** shows flow pattern over a sharp-crested weir. The nappe usually traps a certain amount of air between the lowest nappe surface and the downstream side of the weir. When the downstream water level rises over the weir, the weir is said to be submerged. Eq. (5) is used for estimating the specific energy of water over the sharp-crested weir:

$$E = a + H + \frac{V^2}{2g} \tag{5}$$

where g is the acceleration of gravity = 9.81 m/s^2, a is the height of the weir above the channel bed (m), H is the height of the water surface above the weir crest (m), V is the upstream flow velocity (m/s) and E is the specific energy (m).

Considering a streamline from a point in the upstream flow to a point in the plane of weir on the assumption of uniform velocity in the upstream of flow, the relationship between discharge and water level over sharp-crested weir is presented in Eq. (6):

$$Q = \frac{2}{3} C_d \, B \, \sqrt{2g} \, h^{3/2} \tag{6}$$

where Q is the flow discharge (m^3/s), C_d is the discharge coefficient, B is the width of weir and h is the head over the weir (m).

3.1.3 V-notch sharp-crested weir

V-notch sharp-crested weir is an upright tinning structure placed perpendicular to the base of a horizontal channel. The weir is a common discharge measuring

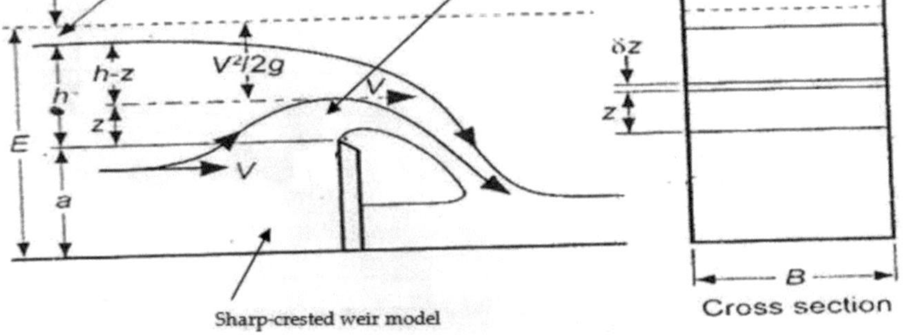

Figure 2.
Flow pattern over a sharp crested weir, where a is the height of the weir crest above the flume channel bed (m); L is the length of the weir (m); V is the velocity of moving water (m/s); y is the depth of water before the critical stage (m); yc is the critical depth (m); Vc is the critical velocity (m/s); H is the head of water above the weir (m); and h is the head of water above the weir at upstream of weir section (m).

device applicable to a various degree of streamflow. Globally, the weir is also known as Thompson weir. The flow regimes that are observed in the weir are:

- Partially contracted weir in which the contraction at the sides of the weir is partly noticed because of closeness to the walls and bed of the channel.
- Fully contracted weir is a channel whose bed and sides are at far distance from the edges of the weir to give enough approach velocity parallel to surface of the weir. Expression to determine discharge over the weir is presented in Eq. (7):

$$Q = \frac{8}{15}\sqrt{2g}\ \tan\frac{\theta}{2}h^{\frac{5}{2}} \tag{7}$$

where Q is the flow discharge (m^3/s), g is the acceleration of gravity = 9.81 m/s^2, h is the water head above the weir vertex (m) and θ is the vertex angle (°).

3.1.4 *Trapezoidal-shaped (Cipoletti) weir*

Modification of fully contracted sharp-crested weir with a trapezoidal control section is known as Cipoletti weir. The weir crest slopes outward with 1:4 horizontal to vertical inclination. Cipoletti in 1886 observed that an increase in side contraction with corresponding increase in head results in reduction of discharge over the weir. The discharge reduction would be taken over by the control section. This will permit the use of head-discharge equation of a full-width rectangular weir. The equation for the determination of discharge over Cipoletti weir is shown in Eq. (8):

$$Q = 3.367C_dLH^{\frac{3}{2}} \tag{8}$$

where Q is the discharge (m^3/s), C_d is the discharge coefficient, L is the effective length (m) and H is the depth of flow of water over the weir base (m).

3.1.5 *Short-crested weir*

Short-crested weirs have characteristics similar to broad-crested and sharp-crested weirs. Streamlines above the crest of short-crested weirs are curved. Some typical examples of the weirs are weir sill with rectangular control section, V-notch weir sill, triangular profile weir (Crump weir), the flat V-weir, cylindrical-crested weir, streamlined triangular profile weirs and flap gates.

3.1.6 *Crump weir*

In 1952, Crump published details of a weir with a triangular profile which had been developed at the Hydraulic Research Station. This was claimed to give a wide modular range and also to give a more predictable performance under submerged conditions than other long-based weirs [7].

Crump proposed upstream and downstream slopes of 1:2 and 1:5, respectively, which were based on sound principle. The upstream slope was designed so that sediment build-up would not reach the crest. The downstream slope was shallow enough to permit a hydraulic jump to form on the weir under modular flow condition, thus providing an integral energy dissipater; also under submerged conditions, losses are not too high and the afflux is minimized. The equation for the determination of discharge over Crump weir is shown in Eq. (9):

$$Q = C_d B g^{1/2} h^{3/2} \tag{9}$$

where Q is the flow discharge (m^3/s), C_d is the coefficient of discharge, B is the width of weir (m), g is the acceleration of gravity (m/s^2) and h is the head above the weir (m).

3.1.7 Ogee weir

This is a special type of weir, generally used as a spillway of a dam [20]. The crest of an Ogee weir rises up to a height of 0.115 H and falls in a parabolic form. The shape of water over an Ogee weir is similar to the shape of the lower of a sharp-crested weir. The expression for determining discharge over an Ogee weir is presented in Eq. (10). An Ogee spillway with a fixed-width curvature can pass more flow. Hence, under low hydraulic heads, it is considered as an economical viable structure [21]:

$$Q = \frac{2}{3} C_d L \sqrt{2g} H^{3/2} \tag{10}$$

where Q is the discharge (m^3/s), C_d is the discharge coefficient, L is the length of the weir (m) and H is the depth of flow of water above the weir crest on the upstream side (m).

3.2 Flumes

Flume is a hydraulic device that can be used to constrict flow scenarios in an open channel for the purpose of measurement. In a broad meaning, a flume can be described as an artificial prismatic open channel in the laboratory. It can be used to simulate various hydraulic parameters in an open channel flow such as: depth and corresponding discharge [4]. Flumes can be generally described as hydraulic structure which water flows through. They can be tailored to a greater range of flows and require less head loss.

The traditional flume is used to measure discharge in agricultural systems. Flumes are designed in order to produce a critical depth in the flume throat and thereby creating a direct relationship between water depth and flow rate. In practice, test data or derivation of empirical relationship based on field research is usually employed in determining the relationship between the water depth and the flow rate.

3.2.1 Accuracy and advantages of the flume

Flumes are self-cleaning due to the fact that the velocity of flow through a flume is usually high and also the absence of no obstruction across the channel. It is also possible to operate with a very smaller head loss which cannot be achieved with a similar weir structure, and this makes flume to be adopted in many areas where the available head is limited. Traditionally, flume is used in measuring flow in agricultural systems and it requires low maintenance cost [22]. It has capacity to measure higher flow rates than a comparably sized weir.

In terms of accuracy, it is possible to obtain an accuracy within ±2–5% (for the flume itself) with overall system accuracy for a typical installation being ±10% when all factors are considered [22].

3.2.2 Classifications of flumes

Flumes are generally classified into two major categories. These are long-throated and short-throated flumes. Examples of long-throated flumes are rectangular (Venturi), trapezoidal and U-shaped flume. Examples of short-throated flumes are Parshall flumes and H-flumes, throatless flume with rounded transition and throatless flume with broken plane transition (cutthroat flumes). When compared with weirs, long-throated flumes are similar to broad-crested weirs (parallel streamlines), while short-throated flumes behave like short-crested weirs (curved streamlines).

3.2.3 Discharge measurement using flumes

Flumes are specially configured open channel structures which control the velocity, resulting in flow changes and in water level. The streamflow through a flume is estimated by measuring depth of water in the flume at a specified location, based on the flume configuration. Generally, flumes are employed to determine discharge where weirs are not useful. They are commonly useful in measuring field runoff when streamflows during storms can be channelled and conveyed across the device. Flumes are very suitable in measuring streamflows containing sediment because the increased velocity through the flume tends to make itself cleaning [7]. The discharge over a flume is determined using Eq. (11):

$$Q = kH_a^n \tag{11}$$

where Q is the flow rate (m^3/s), k is the flume discharge constant (varies according to flume size), H_a is the depth at the site of measurement (m) and n is the discharge exponent (depends on flume size).

3.3 Orifices

Orifice is a small opening provided at the side or bottom of a tank through which liquid flows out. It is used to measure the flow of a fluid from a tank or reservoir. The cross section of the opening may be square, rectangular, triangular or circular. The stream or liquid coming out of an orifice is called a jet. The difference between orifice and notch is size; orifice is smaller than the notch. The quantity of flow measures by orifice is smaller than the notch [22]. Orifices are classified based on size, shape, discharge conditions and shape of the upstream edge. The expression for estimating the amount of flow through an orifice is shown in Eq. (12):

$$Q = AC_d\sqrt{2gh} \tag{12}$$

where Q is the discharge (m^3/s), C_d is the discharge coefficient, A is the area of orifice (m), h is the depth of flow of water over an orifice (m) and g is the gravitational acceleration = 9.81 m/s^2.

3.3.1 Rectangular notch

A notch is an opening in the side of a tank or a vessel in such a way that the liquid surface is below the top edge of the opening. It is a device used to measure the rate of flow of liquid through a tank or a small channel. Notch is a thin structure, usually made of a metallic plate [20]. Flow pattern over a rectangular notch is similar to flow over rectangular and Ogee weirs. The expression for calculating flow through a rectangular notch is presented in Eq. (13):

$$Q = \frac{2}{3} C_d L \sqrt{2g} H^{3/2} \tag{13}$$

where Q is the discharge (m^3/s), C_d is the discharge coefficient, L is the length of the weir (m) and H is the depth of flow of water in the notch (m).

4. Conclusion

Selecting the most appropriate hydraulic structure and the optimal design of its dimension is very critical to the accuracy and quality of streamflow measurement. Therefore, it is incumbent on the designer to balance its choice based on the characteristics of the structures, field constraints and human factors as dictated by the water management in the area. The characteristics of different hydraulic structures for streamflow measurement are embedded in a number of properties such as:

i. Range of measurement is determined by the shape and width of the crest.

ii. Head loss required by hydraulic structures with a high discharge capacity usually has a high coefficient of discharge and vice versa.

iii. Accuracy of measurement is determined by reliability of the calibration. Sharp-crested weir (V-notch) has the highest accuracy ($\pm$2%) when measuring small discharge in an open channel.

iv. Possibilities of regulating discharge over hydraulic devices have been developed to perform more than one function such as streamflow measurement and flood regulation. Other structures such as flumes are designed exclusively to have a fixed crest.

Author details

Adeniyi Ganiyu Adeogun[1]* and Apalando Abdulrasaq Mohammed[2]

1 Department of Civil Engineering, Kwara State University, Malete, Kwara State, Nigeria

2 National Centre for Hydropower Research and Development, University of Ilorin, Ilorin, Nigeria

*Address all correspondence to: adeniyi.adeogun@kwasu.edu.ng

References

[1] Negrel J, Kosuth P, Bercher N. Estimating river discharge from earth observation measurements of river surface hydraulic variables. Hydrology and Earth System Sciences. 2011;**15**: 2049-2058

[2] Goodarzi E, Farhoudi J, Shokri N. Flow characteristics of rectangular broad-crested weirs with sloped upstream face. Journal of Hydrology and Hydromechanics. 2012;**60**(2): 87-100

[3] Dabral PP, Pandey PK, Kumar T, Chakraborty S. Determination of discharge coefficient and head-discharge relationships of different hydraulic structures. Journal of Indian Water Resources Society. 2014;**34**(1): 40-52

[4] Mohammed AA, Balogun OS. Understanding hydraulic models through laboratory investigations of sharp-crested and broad-crested weirs. In: Proceeding of International Civil Engineering Conference Organized by Nigerian Institution of Civil Engineers; 2015. pp. 60-69

[5] Sadhan A. Journal of Fluid Mechanics. 2006;**20**:10-18

[6] Meals DW, Dressing SA. Surface water flow measurement for water quality monitoring projects, Technical Note 3, Developed for U.S. Environmental Protection Agency by Tetra Tech, Inc.; Fairfax, VA; 2008

[7] Chadwick A, Morfett J, Borthwick M. Hydraulics in Civil and Environmental Engineering. 4th ed. Pondicherry, India: Integra Software Services Pvt. Limited; 2004

[8] Afshar H, Hoseini SH. Experimental and 3-D numerical simulation of flow over a rectangular broad-crested weir. International Journal of Engineering and Advanced Technology (IJEAT). 2013; **2**(6):214-219

[9] Haun S, Olsen NRB, Feurich R. Numerical modeling of flow over trapezoidal broad-crested weir. Engineering Applications of Fluid Mechanics. 2011;**5**(3):397-405

[10] Hoseini SH, Afshar H. Flow over a broad-crested weir in subcritical flow conditions physical study. Journal of River Engineering. 2014;**2**(1):1-6

[11] Salmasi F, Poorescandar S, Dalir AH, Zadeh DF. Discharge relations for rectangular broad-crested weirs. The Journal of Agricultural Science. 2011;**17**: 324-336

[12] Jalil SA, Ibrahim AS, Jafer RA. Surface roughness effects on discharge coefficient of broad crested weir. Research Journal of Applied Sciences, Engineering and Technology. 2014; **7**(24):5227-5233

[13] Al-Hashimi SAMM, Madhloom HM, Khalaf RM, Nahi TN, Al-Ansari NA. Flow over broad crested weirs: Comparison of 2D and 3D models. Journal of Civil Engineering and Architecture. 2017;**11**:769-779

[14] Arvanaghi H, Oskuei NN. Sharp-crested weir discharge coefficient. Journal of Civil Engineering and Urbanism. 2013;**3**(3):87-91

[15] Farhoudi J, Alami HS. Slope effect on discharge efficiency in rectangular broad crested weir with sloped upstream face. International Journal of Civil Engineering. 2005;**3**(1):58-65

[16] Arvanaghi H, Naderi V, Azimi V, Salmasi F. Determination of discharge coefficient in inclined rectangular sharp crested weirs using experimental and

numerical simulation. Journal of Current Research in Science. 2014;**2**(3): 401-406

[17] Abd El-Hady Rady RM. 2D-3D modeling of flow over sharp-crested weirs. Journal of Applied Sciences Research. 2011;**7**(12):2495-2505

[18] Naderi V, Nasrabadi MS, Arvanaghi H. Effect of height of sharp-crested weir on discharge coefficient. International Journal of Basic Sciences & Applied Research. 2014;**3**(6):325-330

[19] Adkins GB. Flow measurement devices, Utah Division of Water Right. https://www.waterrights.utah.gov/distinfo/measurement_devices.pdf. 2006

[20] Upadhyay AK. Fluid Mechanics (Hydraulics). New Delhi, India: S. K. Kataria and Sons; 2010

[21] SheikhKazemi J, Saneie M, Moghaddam MA. Experimentally determination of discharge coefficients of Ogee spillway under axial arc condition with convergent lateral walls. Journal of Hydraulic Structures. 2016; **2**(1):27-33

[22] Flumes: retrieved from https://www.openchannelflow.com/flumes. November 25, 2018

Chapter 5

Geotechnical Engineering Applied on Earth and Rock-Fill Dams

Raúl Flores-Berrones and Norma Patricia López-Acosta

Abstract

This chapter presents the importance of geotechnical engineering on the site selection, design, construction, operation, and maintenance of earth-rock dams and earth structures; it emphasizes the geotechnical engineering work related to dam safety during the operation stage. Preliminary geological studies required to select the best dam site are described first. Next, the field and laboratory studies related to the curtain design and dam foundation treatment, as well as geotechnical studies required for the construction, operation, and maintenance of the dam, are discussed. Recent developments in the following three areas are also included: (a) seismic considerations for the design, construction, and maintenance of earth dams; (b) importance of water flow control through the dam embankment and dam foundation, required to avoid internal soil erosion and excessive pore pressure; and (c) dam safety in Mexico and around the world. A case history of a recent failure is used for illustration purposes. In this example, design and construction shortcomings resulted in serious damages on an earth dam. Conclusions and recommendations related to this topic are presented at the end of this chapter.

Keywords: earth-rock dams, dam safety, geotechnical engineering, hydraulic works, soil erosion

1. Background

As is well known, the construction of earth dams dates back many centuries, whose main objective has been the storage of water and flood control.

Due to many reasons, mainly social and environmental, the construction of new dams has led to the operation and conservation of existing structures. This fact has caused geotechnical engineering to be involved in the dam safety management; therefore, it is essential to know the principles of analysis and design with which they were constructed.

There are multiple factors that intervene in the selection of the type of dams; but the topography, regional geology, the availability of construction materials, seismicity of the area, the hydrology of river basin, the environment and geotechnical conditions of the reservoir, and the curtain site are the most important [1]. Thus, the type and classification of the dam dictate the magnitude of the previous and definitive studies to execute the work [2, 3].

IntechOpen

2. Geotechnical engineering in preliminary studies

Geological studies play a key role in the early stages of analysis. Its objective is to know the quality and characteristics of the materials, mainly of the areas where the dam is to be constructed. However, these studies must continue at the preliminary stage, project, and throughout the construction of the work.

Some examples of geological investigations can be geological and geotechnical cartography, exploratory excavations (trenches, galleries, or tunnels), and identification of rock mass discontinuities and their geomechanical classification (e.g., [4–6]).

In the case of soil mechanics studies, it is advisable to include the following activities:

- Collection of information: geologic maps, previous studies, aerial photographs and satellite images, etc.

- Recognition visit: preliminary identification and classification of soils and, in particular, its geological origin, visual classification, and soil site characterization.

- Programming of geotechnical boreholes: depth of the explorations, number and type of boreholes, sampling method, etc.

- Measurement of in situ properties: shear strength, permeability, shear modulus, and instrumentation

- Laboratory tests: index properties, mechanical properties, and compaction tests

Regarding rock mechanics analysis, it is advisable to carry out studies of the rock mass, for example, origin and degree of weathering of the rocks, rock quality, classification, characteristics of the joints, location of faults or old landslides, permeability, compressibility, deformability, shear strength, and susceptibility of the rock to the change of its properties due to wetting.

3. Geotechnical engineering in design

At the design stage, there are two conditions to keep in mind: safety and economy. However, the selection of the type of dam depends on geotechnical aspects such as (a) type, quality, characteristics, and location of the materials for its construction; (b) characteristics of the foundation at the dam site; and (c) stability of slopes in the reservoir and embankments of the dam. In addition, the dam axis depends on the characteristics and geotechnical properties of the foundation.

Concerning the characteristics of the materials for the construction of earth dams, those that are dispersive, collapsible, or susceptible to piping should not be used. In addition, for the design of this type of dams, the mechanical properties of earth materials must be taken into account, for example, shear strength, compressibility, and permeability. If the dam is located in a seismic zone, the dynamic shear modulus and damping of the materials must also be obtained.

Regarding the compressibility and shear strength of the foundation materials, the short- and long-term settlements of the dam have to be analyzed. As for the volume of water that passes through the foundation, the type of treatment and alternatives to reduce the flow will be defined by the permeability of the materials. Similarly, when the dam is located in a seismic zone or an area susceptible to vibrations (e.g., near borrow areas where explosives or related works are used), the susceptibility to the liquefaction phenomenon must be evaluated.

The compacted layers thickness, the inclination angle, the materials, and construction procedures of the dam embankments are governing by the upstream and downstream slope stability analyses. These analyses must consider different loading conditions: full dam, empty dam, rapid filling, rapid drawdown, seismic effect, and so on. Additionally, the susceptibility to piping must be analyzed, which is why it is important to consider the design and installation of filters and transition zones.

Finally, in the economic aspect, transportation and the treatment processes for the materials directly influence the total cost of the work.

4. Geotechnical engineering in construction

The participation of the geotechnical engineer is indispensable in the different stages of the construction of dams, in particular, to verify that the materials and activities correspond to the design and planning of the work.

Special care should be taken in the construction of structures that cross the dam, for example, the spillways and intake structure, among others. The inadequate control of the compaction and characteristics of the materials in the periphery of these elements can cause the phenomenon of internal erosion and piping. In fact, it is in these areas that erosion and piping processes begin. Both situations have caused the failure of numerous earth dams. A particular case is the Teton Dam that [7] described.

It is also important to control the grain-size distribution specifications of the filter and drainage areas, especially when they are placed on the site. There are experiences of plugging and malfunctioning of filters, precisely because of the segregation of soil particles or because of the lack of compliance in their characteristic grain-size.

Likewise, in the construction process, it is important to implement the dam to evaluate its behavior as the construction progresses. The above will allow taking corrective measures before any eventuality, for example, reduce the speed of construction and build berms on the slopes to reduce the stresses acting at the bottom of them, among others.

5. Geotechnical engineering in operation and maintenance

The presence of the geotechnical engineer in the operation and maintenance of a dam helps to solve any eventuality. In addition, its presence supports quick decision-making to avoid damage or failure in the behavior of the dam and the foundation.

According to the life of the dams, the aging of the materials is an important aspect of their safety. This effect causes the earth materials to undergo a change in their structure, composition, and properties, which are due to the microbiological activity of chemical or mechanical processes that have during the useful life of the work. As a result, these changes cause the softening, loss or gain of rigidity or strength of the material, alteration of its hydraulic conductivity, and so on. Aging of the materials can also cause the clogging of drains and filters. Mitchell [8] described an example of a dam affected by age.

It is vital to establish a dam security system that evaluates and prioritizes those that require immediate attention. In this sense, geotechnical engineering can contribute significantly to the design of mechanisms to assess their condition and define the hazard level for their repair or rehabilitation. In this regard, several countries have developed techniques for this essential part of the life of dams [9, 10].

6. Seismic considerations in dams

The damage or effect of an earthquake on earth and rock-fill dams depends on the geology of the site, soil conditions, stability of the foundation, characteristics of the slopes, and slopes in the reservoir, as well as the seismo-dynamic considerations that are taken into account in the analysis of complementary structures (spillways, intake structures, etc.).

The seismic effects on dams usually result in dam settlement, landslides in the area of the reservoir, or embankments. Such effects also appear as cracks in the structures that complement the dam, which can cause excessive water seepage through the dam and soil erosion. In addition, the earthquake can cause soil liquefaction if dam is constructed on low density or uncompacted soil layers [11, 12].

6.1 Analysis methods

There are several methods of analysis for the evaluation of seismic effects in dams; however, they can be summarized in two stages of main interest:

1. Establish the seismic environment under which the dam does not fail. In this case, the ICOLD [13] in its Bulletin 72 recommends considering the design earthquake for two levels of severity: maximum credible earthquake and operating basis earthquake.

2. Define the analysis methodology according to the magnitude and relevance of the dam, as well as the design stage (preliminary or final) and the risk involved in the failure of the dam.

Some traditional methods of analysis for seismic design in dams can be pseudostatic [14, 15] and finite element. **Table 1** summarizes the general aspects of each of them.

The information required by the above methods is summarized in the following points:

a. *Seismicity of the area*. Acceleration and maximum velocity of the terrain, spectral acceleration, magnitude, epicentral distance, seismic sources, return period, and accelerogram representative of the area.

b. *Geological aspects*. The occurrence of superficial faults, induction of seismicity due to the effect of the first filling of the reservoir, and liquefaction of sands in the foundation and/or body of the dam must be foreseen, in addition to the existence of landslides in the area of the reservoir or the dam.

c. *Material properties*. The G modulus (obtained from the dynamic shear stress-strain curve) and hysteretic damping, D.

d. In the case of clays, the dynamic effect of the shear strength with respect to the resistance in static conditions must be taken into account.

e. In the case of sands, susceptibility to liquefaction should be considered according to their relative density.

6.2 Inspection after an earthquake occurred

In dams located in seismic zones, there must be a guide to carry out immediate and subsequent inspections. Both inspections should focus on the dam, the

Generalities	Method			
	(1)	(2)	(3)	(4)
Consider a rigid sliding wedge	X	X		
Consider a single dynamic pulse	X			
The resistance of the soil is assumed constant during the occurrence of the earthquake	X	X	X	
The coefficient of kinetic friction remains constant and equal to static friction	X	X	X	X
The creep acceleration is constant throughout the sliding surface	X	X	X	X
It assumes a single fault surface	X	X	X	
Does not take into account superior vibration modes	X	X	X	
Does not consider the three components of the earthquake	X	X	X	X

Note: (1) Pseudostatic, (2) [15]; (3) [14]; (4) 2D finite element

Table 1.
Overview of traditional methods for seismic analysis of dams.

embedments, the foundation, the area of the reservoir, and the auxiliary works (inlet and outlet channels, spillways, intake structures, tunnels, and electromechanical equipment, among others).

In the case of immediate inspections, it is necessary to report any damage such as landslides, settlements, cracks, and groundwater seepage that did not exist before the earthquake. Similarly, it will be relevant to analyze the information recorded by the instrumentation, mainly to know the condition and behavior of the elements of the dam. The immediate inspection must be maintained for a period of not less than 48 hours. Thus, those responsible may establish emergency strategies if there is a potential risk of failure.

On the other hand, in the subsequent inspections, groups of engineers who are familiar with the project and the construction of the dam must be involved. In this way, the magnitude of the damage and the risk that the work represents are evaluated. Unquestionably, the purpose of the subsequent inspection will be to determine the forms and possible causes of failure. In particular, the inspection should focus on the condition of the foundation, for example, differential settlements, landslides, excessive pore water pressures, groundwater seepage, etc.

It should be added that the North American Great Dams Committee 1986 proposes a format for the inspection of these works. Likewise, [16] mention some recommendations for the inspection of these structures.

7. Relevance of seepage control

There are three main causes related to dam failure due to groundwater movement: (1) soil piping, (2) uplift, and (3) excessive water seepage through the dam.

Some typical measures to solve these problems are (a) adequate selection of materials for construction. (b) reduction of water seepage through a design that considers the geological conditions of the site and the permeability of the materials, (c) control of the compaction of the material and other construction procedures, (d) definition of transition zones between materials of different granulometries (filters), and (e) construction of relief wells that reduce and control pore water pressures.

7.1 Soil piping

Piping is a consequence of the seepage forces caused by the groundwater movement. This phenomenon is produced by the removal and dragging of soil solid particles. Piping occurs when the resistant forces of the soil are less than seepage forces. Resistant forces depend on several factors, but the cohesion, binding, and weight of the solid particles are the most important. Thus, for example, in earth dams, the filters located upstream and downstream of the dam help to combat the piping phenomenon [17]. **Table 2** relates the resistance to piping for different types of soil. In summary, the soils most susceptible to piping are the fine sands that are poorly compacted, whereas the soils with greater resistance are the clays of high plasticity.

The problem of piping can start in any crack caused by differential settlements of the dam, earthquakes, tension cracks, holes left by roots and rotten tree trunks, and, even, by holes or burrows excavated by animals. Frequently, soil piping occurs between contacts of rigid structural members of the dam and loose or poorly compacted materials.

7.2 Seepage forces

The seepage forces intervene in the stability of the slopes of the dam. Among the most critical conditions to which a dam may be subject during its useful life are (a) rapid filling of the reservoir, (b) steady-state groundwater flow with the normal water level, and (c) rapid drawdown of the reservoir. Before this, whatever the condition of analysis, it is necessary to draw a flownet or establish some analytical or numerical procedure to determine the pore water pressures and seepage forces on the failure surface of the slope of the dam.

7.3 Seepage control measures

The reduction of groundwater flow and exit hydraulic gradients is achieved with several procedures. Some examples are cutoff walls, grout curtains, waterproofing membranes or full face waterproofing to the face upstream of the dam, construction of impermeable cores at the dam, partial or total penetration trenches located in permeable zones, etc. Usually, these measures are complemented with the installation of filters and drains [6]. The latter serve as an additional line of defense against water seepage problems. However, filters must meet certain granulometric requirements, which can be found in the methods proposed by Terzaghi [18–23], among others.

7.4 Numerical modeling in groundwater flow analyses

Currently, the increasing development of computers allows the numerical solution of different seepage problems, which, by graphics or analytical methods, are

Resistance	Type of material
High	• Clay with high plasticity, well compacted • Clay with high plasticity, poorly compacted • Well-graded sand-gravel or sand-gravel mixtures packed in the clay of medium plasticity, well compacted
Media	• Fine-graded sand or mixtures of sand-gravel packed in the clay of medium plasticity, poorly compacted • Well-graduated gravel-sand-silt mixtures without cohesion (IP < 6), well compacted
Low	• Gravel-sand-silt mixtures well graduated without cohesion (IP < 6), badly compacted • Fine sands, without very uniform cohesion, well compacted • Fine sands, without very uniform cohesion, poorly compacted

Table 2.
Empirical relationship between the resistance to piping and the type of material [24].

usually difficult to solve [25]. However, numerical modeling makes it easier to study different cases of analysis, for example, heterogeneous and anisotropic groundwater flow, steady-state and transient state flows, as well as the groundwater flow in saturated and unsaturated media. In this area, the methods of finite differences (**Figure 1**) and finite elements (**Figure 2**) are usually the best known. Different publications show the facilities provided by these procedures [26–28].

Numerical techniques have wide advantages over any other analysis procedures. Two of the most relevant are coupled groundwater flow-slope stability analyses and groundwater flow deformation. In both cases the scope is to analyze the pore water pressure, with which more realistic results are obtained from the problem analyzed [29].

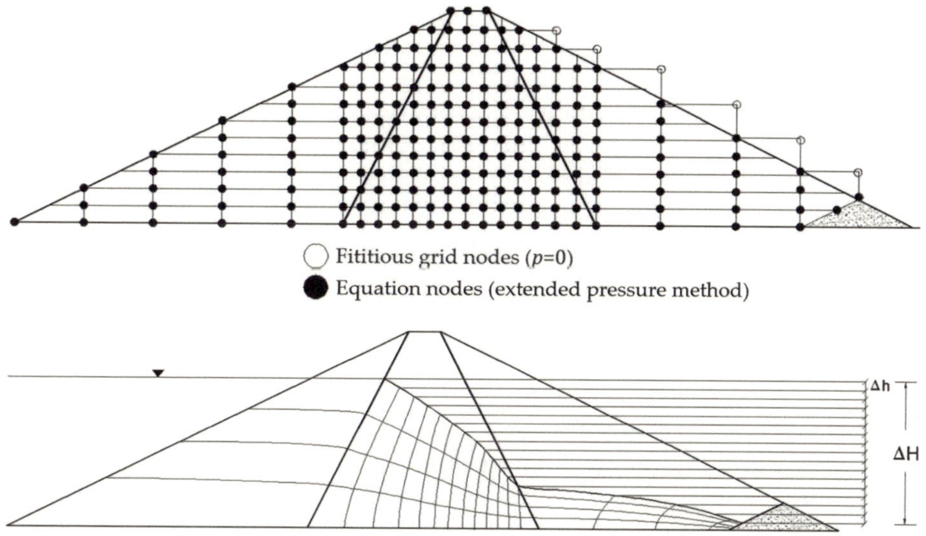

Figure 1.
Solution of a water flow problem by the finite difference method [30].

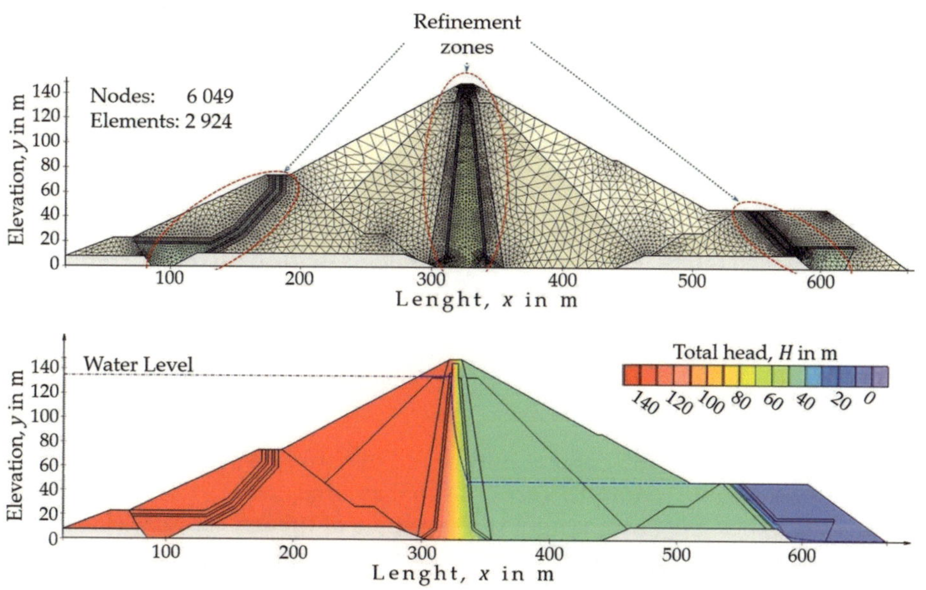

Figure 2.
Solution of a water flow problem by the finite element method [29].

8. The security of dams in Mexico

In Mexico, from 1926, when the National Irrigation Commission was created, the design and construction of several dams were undertaken in order to store water for agricultural irrigation, hydroelectric generation, water supply, flood control, aquaculture, and recreation. These works allowed dam engineering and soil mechanics in Mexico to become the two disciplines with international prestige for Mexican engineering.

The design, construction, and operation of dams in Mexico, especially earth and rock-fill dams, have had a high level of technological development, whose influence has transcended in the international arena. However, until recently, there was no official standard for the safety of dams that would specify in a theoretical and legal framework the minimum requirements to be met in the basic stages of the life of the dams, that is, design, construction, first filling, operation, maintenance, and abandonment.

Recently, the National Water Commission (formerly the National Irrigation Commission) established a multidisciplinary working group made up of experts from various government agencies and research institutions. Its objective was to establish a Mexican Standard, in addition to setting the minimum requirements that must be met in the different life stages of a dam and establishing the solution plans for different emergency scenarios, as well as the management and technical decisions that significantly affect the safety of existing dams in the national territory, and thus reduce risks to people, property, and the environment.

8.1 International context

In the last 30 years, the construction of new dams has decreased in developed countries [9]. Currently, the actions have been aimed at rehabilitation, prolonging the useful life and increasing safety and final closure of those that are not in operation. Therefore, dam safety management programs have focused on:

- Maintain and operate the dam at the security level considered in its design.
- Verify that the dam meets design expectations, and identify possible deviations from safety levels by monitoring its behavior and surveillance.
- Periodically review the design and performance of the dam to identify safety problems and formulate action or remediation measures.
- Establish action plans in case of any incident of the dam.
- Decide which safety issues or existing problems in the dam require immediate attention or can be handled within a framework of the dam safety improvement program.

8.2 Situation in Mexico

In Mexico, there are 5166 dams administered by different agencies. Of the slightly more than 5000 dams, 836 are classified as large dams [31]; most of them with earth and rock-fill curtains built in the middle of the last century. However, there are also a significant number of masonry and concrete curtains; several of them are from the nineteenth century or the beginning of the twentieth century.

A large part of the dams in the country is about to reach their useful life. Others are at an advanced age, that is, between 20 and 50 years old. Seventy-one percent of the dams exceed 20 years old, and on average, they are 36 years old. Therefore, in dams where there is no efficient maintenance, there will be a loss of capacity due to silts, contamination, and possible deterioration of the curtain and its elements.

Another situation that is observed is the change of use for which the dam was originally built. Dams initially intended for irrigation or hydropower generation are now adapted for human consumption. Similarly, projects that were developed for the storage of rainwater are now mixed with wastewater, which attacks normal concrete and reinforcing steels.

8.3 Safety assessment of dams

To know the security status of a dam, it is advisable to carry out annual inspections, every 5 years and whenever an extraordinary event occurs. Botero et al. [9] recommend that the annual inspections seek to know the behavior and operation of the dam in the short term. In inspections every 5 years, a detailed analysis of the condition of the dam must be made and possible corrective actions identified. Finally, in the case of inspections after an extraordinary incident, these should focus on the study of the effects that it could have on the dam. In any case, inspections should consider the following aspects:

- Collect all available information of the dam followed by a visual visit to the site.
- Know the behavior of the dam under normal and extraordinary conditions.
- Review hydrological safety.
- Update weather information.
- Evaluate the stability of the dam and the surrounding structures.
- Identify any problem or incident in the area of the dam and its foundation.
- Evaluate the seismic risk.
- Define the safety factors for the steady-state conditions, rapid drawdown, and earthquake for different reservoir water levels.
- Calculate the stress state and deformations, pore water pressures, groundwater discharge, and free board losses and identify landslides on the reservoir.

Additionally, **Figure 3** shows a flowchart for decision-making in the evaluation of dam safety. In addition, **Table 3** indicates a classification of risk levels in dams.

8.4 Dam removal

With regard to dams that have ceased to operate or are abandoned, the concessionaire must prepare a project that defines the work for putting out of service and the conditions in which the area of influence will remain. In addition, the project must specify the necessary adaptations so that an incident does not occur due to the inappropriate behavior of the remnants of the dam. Therefore, in order to ensure that the removal is carried out correctly, it is necessary to verify that any danger to human lives or important properties has been eliminated.

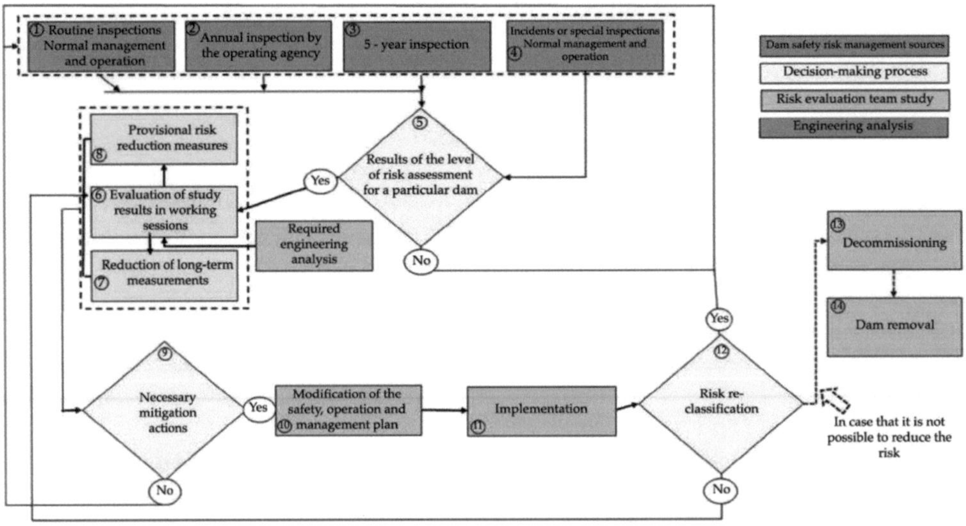

Figure 3.
Flowchart of the proposed method for structures higher than 15 m (modified from [32]).

Risk level	Description
Remote	The physical conditions for the development of a problem are nonexistent or unlikely to occur
Very low	The possibility cannot be taken into account, but there is no convincing evidence that it has ever happened or that there cannot be a situation that could lead to the development of that failure
Low	The causes of the defects are known. Indirect evidence suggests that it is feasible but indicates a low probability of failure
Moderate	The fundamental conditions or defects that can produce the fault are known, and the evidence suggests that it is directly possible
High	There is considerable evidence, direct or indirect, suggesting that such an event has occurred or is likely to happen
Very high	There is direct evidence that the problem is occurring actively or is very likely to happen

Table 3.
Risk levels in dams [9].

It should be noted that the problem of the removal of the dam does not end with the removal of the structures. The final phase consists of determining the impact of sediments on water quality and concentration of pollutants, flood potential, conservation of fish and wildlife, downstream infrastructure, cultural resources, and recreational activities that were carried out in the reservoir and determining if it is necessary to build new structures or mitigation measures.

9. An historical case of an earth dam failure

In this section, a historical case of the failure of a homogeneous earth dam is described. The dam was built in 2009 in San José Iturbide, Guanajuato. The objective of the dam was for the recharge of an aquifer and the supply of drinking troughs by capturing rainwater and runoff from the mountains. The height of the dam is 13.6 m, and the length is 98.5 m. The storage capacity of maximum water level (MWL) is 0.290 hm^3 (Elev. 2199.50 m.a.s.l). The material for the construction consisted of a sandy silt with some gravels and small boulders.

Downstream, approximately 3.7 km away, is the population called El Capulín, whose number of inhabitants exceeds 3300. In addition, it should be clarified that the dam was designed and built in a particular way without the consent of the National Water Commission.

In February 2010, the spillway collapsed during the first filling of the reservoir, for which more than 2000 people from El Capulín population were evacuated. **Figure 4** shows the failure of the dam. The technical visits after the failure showed that the material for the construction was highly erodible and was placed without any compaction procedure. On the other hand, evidence was found that there was no cleanliness in the contact of the dam with the basalt rock. In addition, water seepage was observed in the dam-rock basal contact, and several local type faults were identified on the upstream side of the dam.

9.1 Hydrological review

The results of the hydrological study for maximum avenues with different return periods (including 1 of 10,000 years) showed that, from a return period of 50, the maximum level of the dam was exceeded with hydraulic heads between 0.18 and 1.91 m. As a result, the safety of the structure due to overflow and erosion represented an imminent danger. In fact, the hydrological risk represented by the dam was high.

9.2 Geotechnical considerations

For the geotechnical investigation, six test pit excavations were carried out to obtain soil samples to determine the index and compaction properties of the material. In addition, in situ permeability tests were performed with the Matsuo Akai method. **Figure 5** shows the maximum section of the dam with the explorations made.

Figure 4.
Overflow and failure of the spillway of the La Salitrera dam, in 2010.

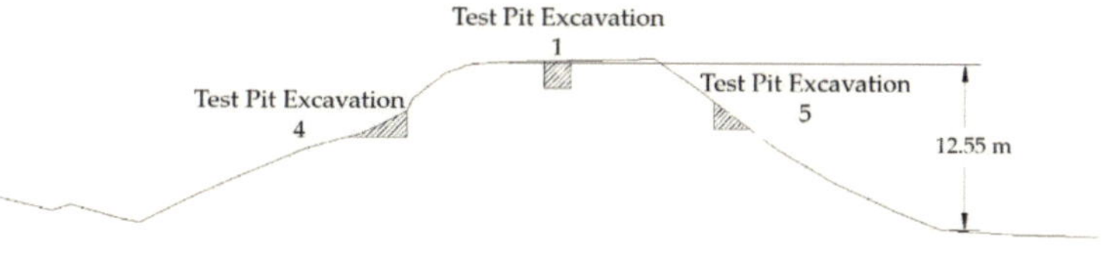

Figure 5.
Maximum section of the dam with the location of test pit excavations and material properties.

Slope stability analyses were carried out with two limit equilibrium methods: Morgenstern-Price and Bishop. Initially, the stability of the dam (upstream and downstream sides) focused on steady-state conditions for water reservoir levels: normal water level (NWL) and maximum water level (MWL). **Table 4** indicates the factors of safety for both conditions. In summary, it is observed that the factors of safety for the slopes of the dam do not satisfy the minimum factor of safety requested by CONAGUA (FoSmin = 1.5). **Figure 6** shows the minimum factor of safety and the failure surface for the downstream side.

On the other hand, the slope stability was analyzed, assuming a rapid drawdown of the reservoir. **Table 5** indicates that the dam slopes have a factor of safety less than 1.2. Therefore, both slopes of the dam are unstable and represent an imminent risk of failure.

Finally, the earthquake stability of the dam was determined. The water level was assumed at NWL, and a return period for the earthquake of 475 years was considered. As shown in **Table 6**, the safety factor for the analysis is greater than 1.0. Therefore, the stability of the dam under a seismic event is not critical. **Table 6** summarizes also the results for the three conditions of operation of the dam: normal, unusual, and extreme.

9.3 Delimitation of flood danger zones

The evaluation of the danger zones by the flood was determined taking into account two factors of main interest: (1) maximum depths and (2) maximum flow velocities. In this case, the hydraulic analyses showed that the channel has a capacity for an avenue with a return period of 10 years. In addition, the estimated maximum water velocities exceed 4 m/s, and the maximum depth of the avenue is greater than 1.5 m.

Flow condition	Slope	Factor of safety		Water level
		Morgenstern-Price	Bishop	
Steady-state	Upstream	3.105	3.108	MWL
Steady-state	Downstream	0.975	0.961	MWL
Steady-state	Upstream	1.815	1.820	NWL
Steady-state	Downstream	1.310	1.306	NWL

Note: MWL = maximum water level

Table 4.
Factors of safety determined with slope stability analysis under steady-state groundwater flow conditions.

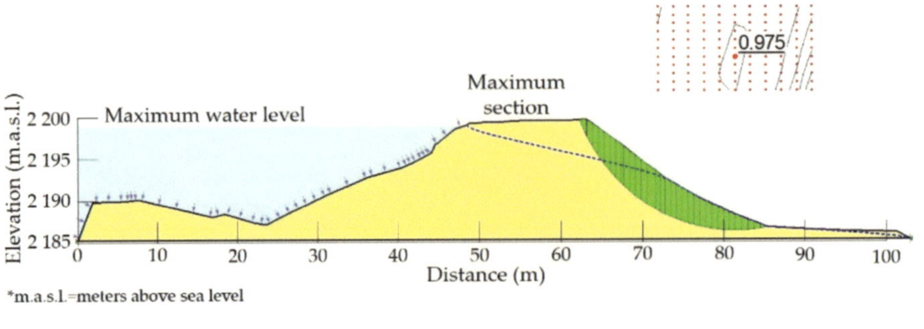

Figure 6.
Minimum factor of safety and critical failure surface in downstream side, steady-state groundwater flow (maximum water level).

9.4 Risk mitigation measures

Three alternative solutions are presented:

1. Construct a cylindrical type spillway (Creager) with a length of 16 m with a water head of 2.75 m and a capacity of 145 m^3/s. These characteristics correspond to a return period of 10,000 years.

2. Cut down the height of the dam by 3 m, and build a 3 m high channel wall from the left embankment; clean and excavate the ignimbrite rock 10 m long and up to the height of 2192 m above sea level, at the right bank, where the spillway was.

3. Remove the dam.

From these three alternatives, alternative 2 is considered the most feasible. As a result, a 10 m long wide crest spillway is proposed. The height of the dam must be reduced by 3 meters to reach a height of 2196.5 m.a.s.l. The modification of the NWL to the level 2192 m.a.s.l. restricts the water level so that it does not represent a risk to the dam. In addition, the modification of the MWL to the level 2195.7 m.a.s.l. will allow the transit of extreme events without affecting the security of the dam.

With regard to the mitigation of flood risk, cleanup work and release of the main channel are recommended. In addition, build the necessary structures such as bridges, culverts, etc. which will allow the free flow of water during extraordinary events.

Flow condition	Slope	Factor of safety		Reservoir water level
		Morgenstern-Price	Bishop	
Rapid drawdown	Upstream	0.838	0.803	MWL to NWL
Rapid drawdown	Downstream	0.976	0.962	MWL to NWL

Note: MWL = maximum water level; NWL = normal water level

Table 5.
Factor of safety determined with the slope stability analysis under rapid drawdown groundwater flow conditions (Section 1).

Operation	Slope	Factor of safety Morgenstern-Price	Flow condition	Reservoir water level
Normal	Upstream	3.11 > 1.50	Steady-state	MWL
	Downstream	0.98 < 1.50	Steady-state	MWL
Unusual	Upstream	0.86 < 1.20	Rapid drawdown	MWL to NWL
	Downstream	0.98 < 1.20	Rapid drawdown	MWL to NWL
Extreme	Upstream	1.72 > 1.00	Steady-state with an earthquake*	NWL

**Note: Earthquake with PGA = 0.45 g and Tr = 475 years.*
MWL = maximum water level; NWL = normal water level

Table 6.
Factors of safety obtained from slope stability analysis under normal, unusual, and extreme operating conditions.

9.5 Additional geotechnical considerations

As can be seen in **Tables 4** and **5**, the stability of the dam in steady-state and rapid drawdown groundwater flow conditions and the factors of safety are less than unity; therefore, the structure is unsafe. Also, taking into consideration the granulometric characteristics of the construction material of the dam, in addition to the photographic evidence of **Figure 4**, the material seems susceptible to internal erosion (piping); therefore, the safety of the dam is not adequate from this point of view.

10. Conclusions

The main conclusions of this work are:

- Geotechnics plays a fundamental role in the proper functioning of earth and rock-fill dams. For this reason, it is essential to perform geotechnical studies from the selection of the dam site and its intervention in the methodology of the design of the embankment and other elements that make up the dam and to monitor the behavior of the entire dam during its useful life.

- The failure of a dam can occur due to a deficient preliminary geotechnical study, a bad design of its embankment or foundation, a poor control in the quality of the construction, or a lack of maintenance in the instrumentation and operation of the dam.

- There are methods nowadays to adequately take into account the forces originated from earthquakes, the groundwater flow, and the problems of internal erosion (piping).

- The inspection visits to each dam should be carried out periodically and, especially, immediately after any extraordinary phenomenon, such as an extreme flood, an earthquake, or any other anomalies not contemplated in its design.

- It is convenient that the inspection visits are carried out by engineers with width experience and that they include the areas of hydraulics, hydrology, geotechnics, structures, and electromechanics.

Acknowledgements

The authors acknowledge and are in particular grateful to José Alfredo Mendoza Promotor for his valuable help in reviewing the format and the edition of this document.

DOI: http://dx.doi.org/10.5772/intechopen.84899

Author details

Raúl Flores-Berrones[1]* and Norma Patricia López-Acosta[2]

1 Mexican Institute of Water Technology, MOR, Mexico

2 Instituto de Ingeniería-UNAM, CDMX, Mexico

*Address all correspondence to: rflores@tlaloc.imta.mx

References

[1] Marsal R, Reséndiz D. Presas de tierra y enrocamiento. México, DF., México: Limusa; 1975

[2] Flores-Berrones R, Vassilev V, Li Liu X. Geotecnia en ingeniería de presas. Jiutepec, Morelos, México: Instituto Mexicano de Tecnología de Presas; 2001

[3] Flores-Berrones R. La geotecnia en ingeniería de presas de tierra y enrocamiento. In: Séptima Conferencia Raúl J. Marsal. Mérida, Yucatán, México: Sociedad Mexicana de Ingeniería Geotécnica; 2016. p. 110

[4] Barton N, Lien R, Lunde J. Engineering classification of rock masses the design of tunnel support. Rock Mechanics. n.d.;**6**:186-236

[5] Bieniawski ZT. Classification in rock engineering applications. In: Proceedings of the 4th International Conference on Rock Mechanics. The Geomechanics; vol. 2. 1979. pp. 41-48

[6] González de Vallejo I. Las clasificaciones geomecánicas para túneles. Ingeotúneles. Madrid, España; 1998

[7] Peck R. Report to U.S. Department of the Interior and State of Idaho Failure of Teton Dam Idaho Falls. Idaho: U.S. Govt. Printing off; 1976

[8] Mitchell JK. The fourth dimension of soil behavior in geotechnical engineering. In: Seventeenth Nabor Carrillo Lecture. Sociedad Mexicana de Mecánica de Suelos; 2004

[9] Botero-Jaramillo E, Ossa-López A, Flores-Berrones R. Strategies for dam safety risk management in Mexico. Tecnología y Ciencias del Agua. 2015;**6**(3):5-14

[10] Ramírez-Reynaga R, Silva FA. Simplified ranking procedure to assess dam safety in México. In: Proceedings, International Conference on Soil Mechanics and Foundation Engineering. Tokyo, Japan; 2005

[11] Casagrande A. Characteristics of cohesionless soils affecting the stability of slopes and earth fills. In: Contribution to Soil Mechanics 1925-1940. 1936. pp. 257-276

[12] Castro G. Liquefaction of sands. Harvard Soil Mechanics Series (No. 81). 1969

[13] ICOLD. Rock fill dams with concrete facing. In: Estate of the Art. International Committee on Large Dams. Bulletin 72; 1989

[14] Makdisi FI, Seed HB. Simplified procedure for estimating dam and embankment Earthquake-induce deformations. Journal of Geotechnical Engineering ASCE. 1978;**104**(7):849-867

[15] Newmark NM. Effects of earthquakes on dams and embankments. Geotechnique. 1965;**15**(2):139-160

[16] Fell R, McGregor P, Stapleton D, Bell G. Geotechnical Engineering of Dams. Balkelma Publishers; 2005

[17] Flores-Berrones R, López-Acosta NP. Chapter 14: "Internal Erosion Due to Water Flow Through Earth Dams and Earth Structures". Part 3: Other Applications in Engineering/ Geoscience. In: Danilo G, Silva S, editors. Soil Erosion Studies. InTech; 2011. pp. 283-306

[18] Foster MA, Fell R. Assessing embankment dam filters that do not satisfy design criteria. Journal of Geotechnical and Geoenvironmental Engineering. 2001;**127**(5):398-407